Physical Plant Equipment Fundamentals

Physical Plant Equipment Fundamentals

Kenneth Lee Petrocelly, RPA, SMA, FMA

LONDON AND NEW YORK

Published 2020 by River Publishers
River Publishers
Alsbjergvej 10, 9260 Gistrup, Denmark
www.riverpublishers.com

Distributed exclusively by Routledge
4 Park Square, Milton Park, Abingdon, Oxon OX14 4RN
605 Third Avenue, New York, NY 10017

First issued in paperback 2023

Library of Congress Cataloging-in-Publication Data

Names: Petrocelly, K. L. (Kenneth Lee), 1946-
Title: Physical plant equipment fundamentals / Kenneth Lee Petrocelly, RPA, SMA, FMA.
Description: Lilburn, GA : Fairmont Press, Inc., [2018] | Includes index.
Identifiers: LCCN 2018012260 | ISBN 0881737968 (alk. paper) | ISBN 9781138616097 (Taylor & Francis distribution : alk. paper) | ISBN 0881737976 (eISBN) | ISBN 9788770222631 (eISBN)
Subjects: LCSH: Factories--Equipment and supplies. | Plant maintenance.
Classification: LCC TH4511 .P47 2018 | DDC 670.028/4--dc23
LC record available at https://lccn.loc.gov/2018012260

Physical plant equipment fundamentals / Kenneth Lee Petrocelly
First published by Fairmont Press in 2018.

Routledge is an imprint of the Taylor & Francis Group, an informa business

Publisher's Note
The publisher has gone to great lengths to ensure the quality of this reprint but points out that some imperfections in the original copies may be apparent.

13: 978-87-7022-949-4 (pbk)
13:978-1-138-61609-7 (hbk)
13:978-87-7022-263-1 (online)
13:978-1-00-315134-0 (ebook master)

Dedication

This work is dedicated to all of the old timers that I've worked with and learned from, who didn't have the benefit of a manual to introduce them to their trades.

Table of Contents

Preface and Introduction

As a young man, I was fortunate to get a head start on my eventual career in the physical plant. Thanks to the great training I received as a boiler technician in the U.S. Navy, I was well prepared for and able to enter civilian life as an operating engineer. As a licensed stationary engineer, my role quickly expanded into electrical and mechanical equipment maintenance, which involved the need to acquire a more in-depth understanding of the equipment that I was operating. Back in "the day," there weren't many schools available to teach the craft, and much of my training was a combination of trial and error, on-the-job training and mentoring from the old timers who had "been there and done that!" As I looked back on my experience (which, by the way, definitely is the best teacher), I often wondered how much easier the transition would have been, had there been a text available that made sense of all that I was confronted with at the many facilities where I worked.

Physical Plant Equipment Fundamentals uncomplicates the care and repair of common plant equipment, providing direction to mechanics and electricians, and furthering their base understanding through referenced examples. It addresses many of the knowledge gaps of the untrained or inexperienced maintenance mechanic, but both fledgling and journeyman-level repair personnel can benefit from the instruction and advice it offers. The book is organized into four sections:

SECTION I: *Essential Plant Systems and Equipment*—provides an inventory of systems and equipment commonly found in most facilities; shares their types, and shows their interiors, parts and process actions.

SECTION II: *Associated Accessories and Mechanisms*—depicts various support accessories, devices and appurtenances associated with the afore-mentioned equipment.

SECTION III: *Operations and Maintenance Support*—addresses items, procedures and processes that can, and often do, assist in their operations and maintenance.

SECTION IV: *Plant Engineering Principles and Norms*—sheds light on the science behind equipment operations and imparts a cursory understanding of its function.

Section I

Essential Plant Systems and Equipment

AIR COMPRESSORS

An air compressor is an air-, water- or oil-cooled machine which is driven by steam, an electric motor, or diesel or gasoline engine that extracts air from the atmosphere (at atmospheric pressure), compresses it (to a higher pressure), and delivers it into a holding tank. When the tank pressure reaches its upper limit, the air compressor shuts off. There, the compressed air is held in the tank until called into use. When the tank pressure reaches its lower limit, the air compressor starts again, re-pressurizing the tank. The most common uses are for pressurizing pneumatic control systems, inflating tires, the operation of pneumatic tools (such as jackhammers) and air brakes. Although air is the most frequently compressed gas, natural gas, oxygen, and nitrogen (among others) are also often compressed. The four basic types of compressors are reciprocating, rotary screw, centrifugal and axial-flow (see Table I-1).

Table I-1

Type	Common Uses
Reciprocating	Refrigeration plants, pneumatic HVAC system control valves, and garage facilities
Rotary	Chemical industry, automotive air conditioning
Centrifugal	Air-conditioning and refrigeration, HVAC, mining and petroleum refining
Axial-flow	Blast furnaces, steel foundries, gas turbine installations, and jet aircraft engines

The reciprocating (or piston-and-cylinder) compressor is the most popular type, and is useful for supplying small amounts of a gas at relatively high pressures. In this type of compressor, a piston is driven within a cylinder. The gas is drawn in through an inlet valve on the suction stroke of the piston and is compressed and driven through another valve on the return stroke. Reciprocating compressors are either single or double acting. In single-acting machines the compression takes place on only one side of the piston, while double acting machines use both sides of the cylinder for compression. Multiple cylinder arrangements are common. See Figure I-1.

Figure I-1
Reciprocating Compressor Showing Pistons

The rotary-screw compressor uses two meshed rotating helical rotors (within a casing) to force the gas into a smaller space. This type of compressor provides smooth, pulse-free gas output with a high output volume. They provide positive displacement compression by matching two helical screws that (when turned) guide air into a chamber whose volume is decreased as the screws turn. See Figure I-2.

Centrifugal compressors consist of a rotating impeller mounted in a casing and revolving at high speed. This causes the gas (which is continuously admitted near the center of rotation) to experience an outward flow and pressure increase, due to centrifugal action. This design type is capable of compressing large volumes of gas to moderate pressures and smoothly discharges the compressed gas. See Figure I-3.

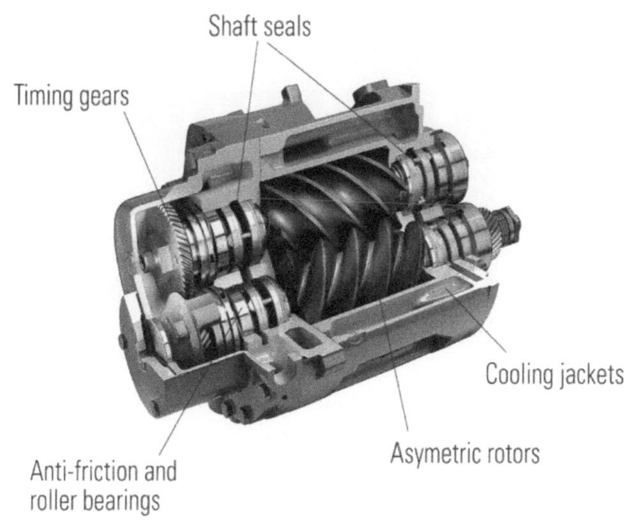

Figure I-2
Rotary Screw Compressor Internals

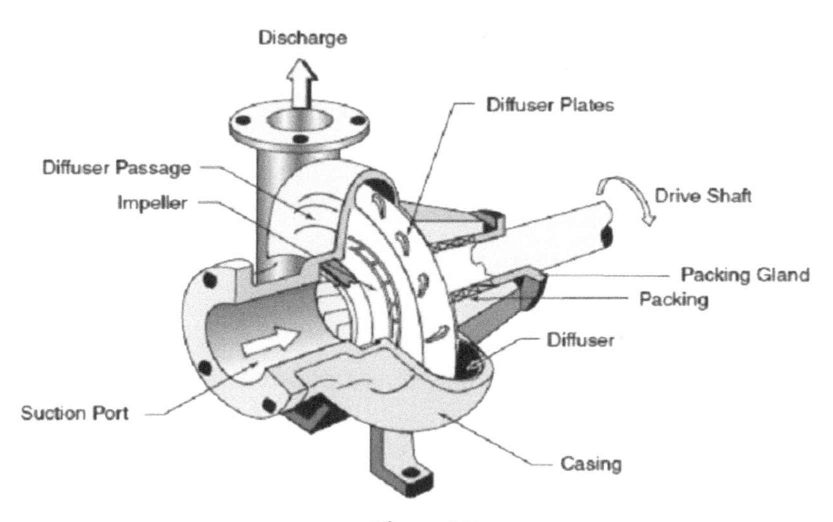

Figure I-3
Centrifugal Compressor Parts

In axial-flow compressors, the gas flows over a set of airfoils spinning on a shaft in a tapered tube. These draw the gas in at one end, compress it, and output it at the other. Axial-flow compressors are used in jet aircraft engines and gas turbines. See Figure I-4.

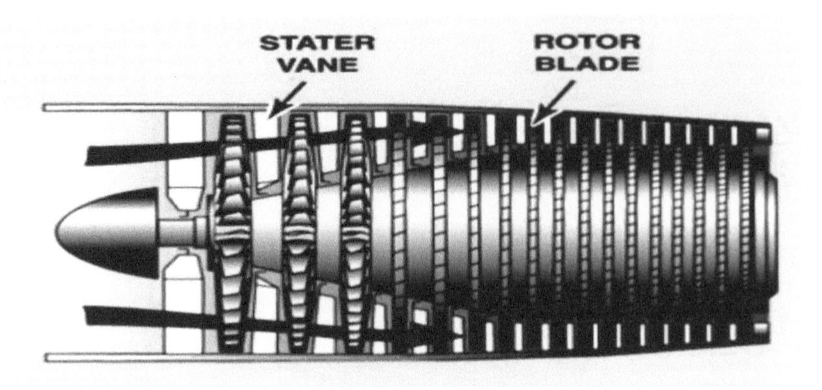

Figure I-4
Axial Flow Compressor

Compressors can be classified according to the pressure they develop. Low pressure compressors have a discharge pressure of 150 psi (pounds per square inch) or less. Medium pressure compressors have a discharge pressure of 151 psi to 1,000 psi. High pressure compressors have a discharge pressure above 1,000 psi. Air compressors require some means for disposing of adiabatic (waste) heat. **Note**: Adiabatic heat is (unwanted) heat created due to a change in the temperature of a material without addition of heat. That happens when the pressure of a gas is increased. The cooling is usually afforded by some form of air-, water- or oil-cooling that is, in turn, air or water-cooled. See Figure I-5.

BOILERS

Boilers are classified by material composition, manufacturer, design, application, tube configuration, number of passes, furnace type, fuel burned, horsepower rating, heating surface, fluid medium... well, you get the idea. As an inspector years ago, I visited the interior of boilers running the gamut from miniature flueless units used to generate steam for pressing garments in a dry cleaning shop, to 20-story once-through, tri-fuel utility boilers operating in excess of 4500 psi. I even inspected a locomotive or two. And through it all was manifest a common thread: If you don't take care of them, they won't take care of you! The boilers most often associated with commercial and institutional heating plants are the package fire-tube, water-tube and sectional types.

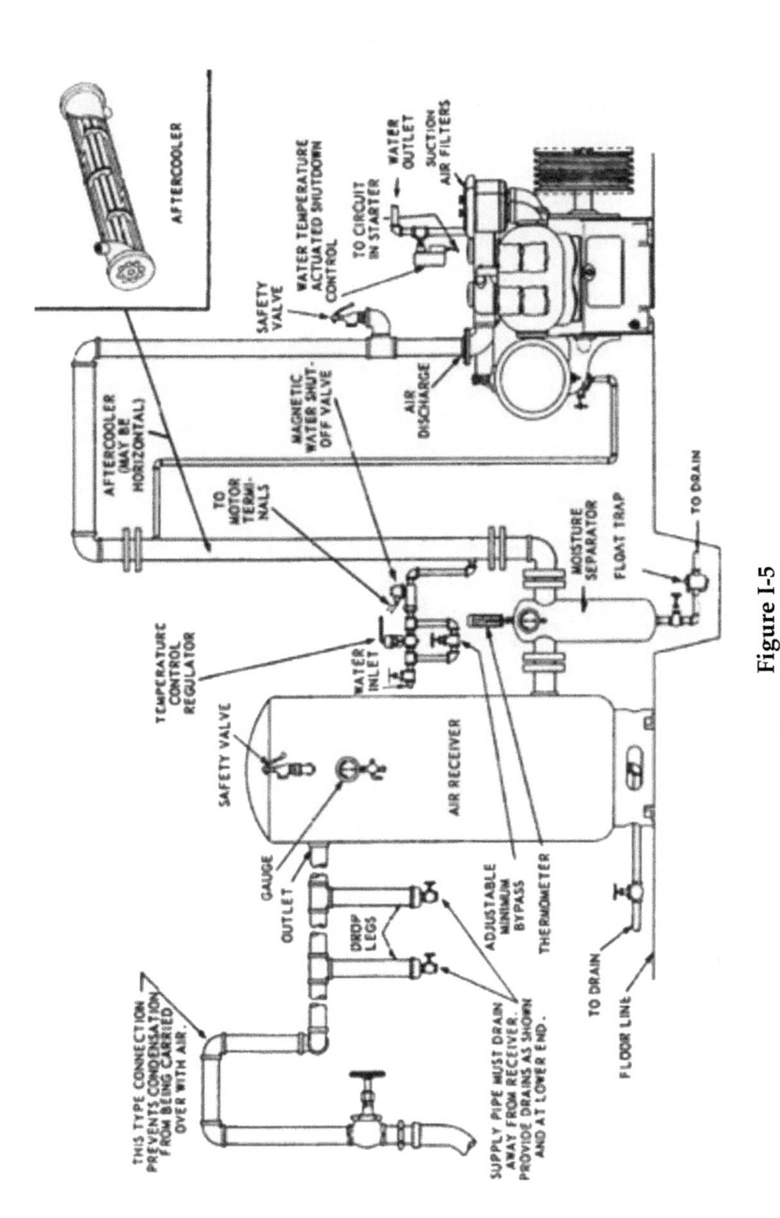

Figure I-5
Main Components of an Air Compressing Plant

Fire Tube

Fire tube boilers are made of steel and manufactured in sizes up to 15,000 pounds of steam per hour with maximum allowable working pressures up to 250 pounds per square inch. Physically, they come in a variety of sizes and configurations and are principally used for heating systems. Low pressure units are limited to 15 psig for steam service and a maximum of 160 psig and 250 degrees Fahrenheit for water service. Their large water storage capacities are useful in controlling the affects of sudden load fluctuations, but considerable time can be lost in bringing them up to operating pressure from a cold start. Heat exchange is accomplished by passing hot gasses through metal tubes which are surrounded by water. See Figure I-6.

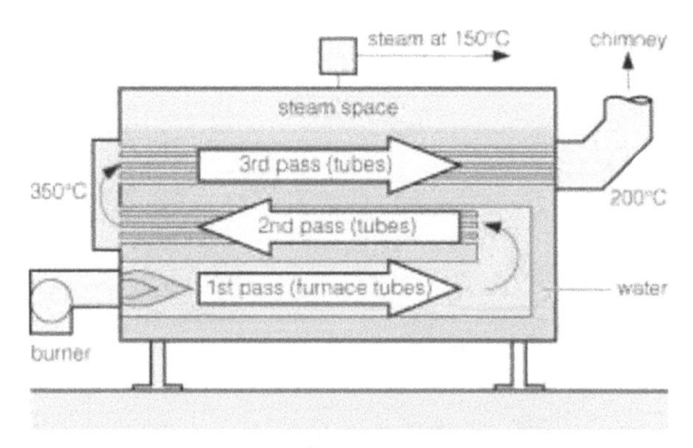

Figure I-6
Hot Gas Passes in a Fire Tube Boiler

Water Tube

Water tube boilers are made of steel and used when large steam generating capacities and high pressures are required. As a rule, they have better efficiency ratings than do fire tube boilers but cost more to install initially. Heat exchange takes place in the generating tube bank, wherein water is passed through metal tubes which are surrounded by hot gasses from the heat source. See Figure I-7.

Sectional

Sectional boilers are most often made of cast iron and used primarily for low pressure applications of 15 psig for steam service and 30 psig for water service. Operating and maintenance costs are much lower than for fire tube and water tube boilers, but initial erection costs can be high because fit-up and assembly are usually performed on the premises. Heat is transferred in this unit by directing hot gases between cast iron sections in which water is contained. See Figure I-8.

Though uniquely configured, each utilizes the same principles of heat transfer and, depending on design, can be used for either steam or hot water service in high or low pressure applications. Each also requires much the same attention to keep them in optimum operating condition. To ensure their safe and efficient operation, it's imperative

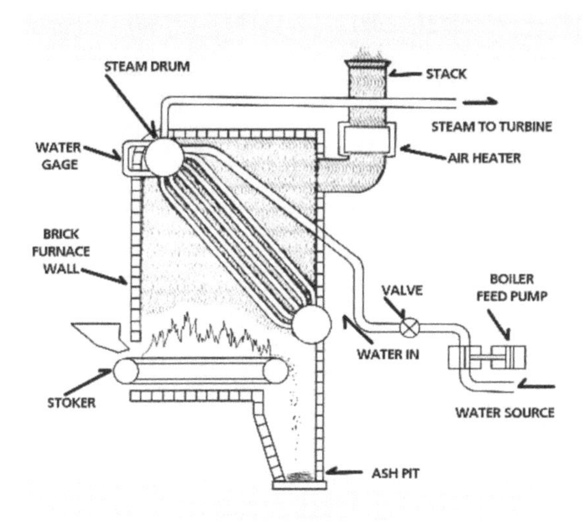

Figure I-7
Water Tube Boiler

Figure I-8
Interior of a Cast Iron Sectional Boiler

that boiler operators become thoroughly familiar with all of their boilers appurtenances and how they function. Figure I-9 is representative of the devices and controls found on most boilers.

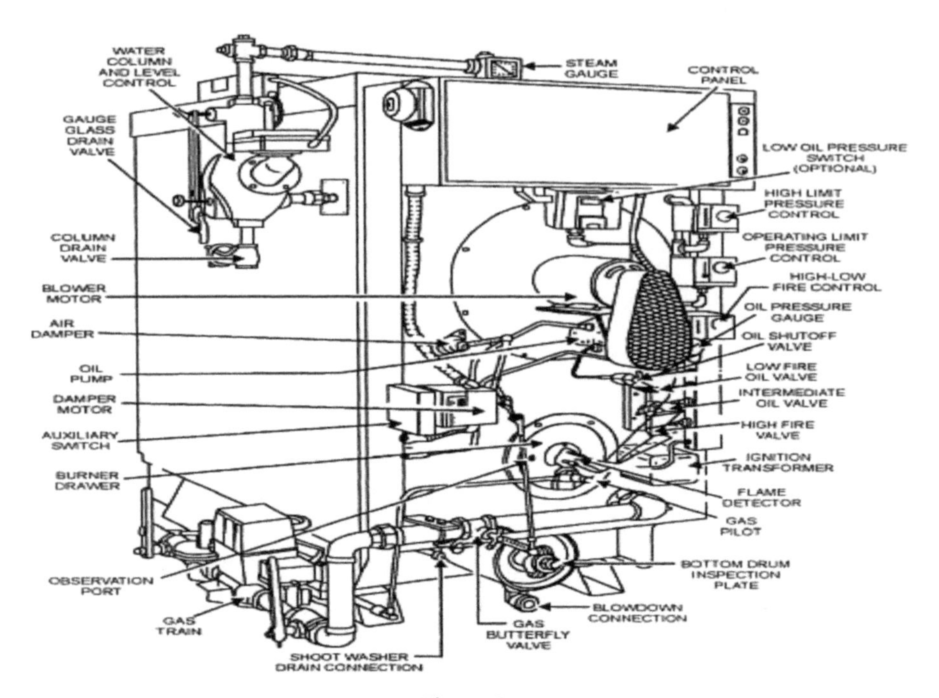

Figure I-9
Common Boiler Appurtenances

The most common boiler failure is from overheating due to operating the unit while it is suffering from a "low water" condition (water level in the boiler is below the normal operating level). The two controls most closely associated with the prevention of this condition are the feedwater regulator, which causes water to be added to the boiler during operation to maintain the normal operating level and the low water fuel cut-off, which shuts the fuel supply off to the burner when the boiler water level drops below a predetermined safe level. Both of these important devices should be tested frequently in accordance with the boiler and control manufacturer's recommendations and as required by your inspector. Basically, there are two ways of testing these controls. One is by the quick drain method, and the other by the slow drain method.

Slow Drain Test—The slow drain test simulates a gradually developing low-water condition and is applicable to all types of low-water fuel cutoff and alarm devices on steam boilers. The test should be made with the burner in operation. Shut off the condensate return and feed supply so that the boiler will not receive any replacement water, and permit the water level to drop. The test can be expedited by opening the blowdown valves. The gage glass should be closely watched and the water level noted at the moment of cutoff and/or sounding of the alarm. If the device fails to function and shut off the valve at the proper level, close the blowdown valves immediately, and restore the water level to normal. **Note**: Do not operate the boiler unattended until the cause of the malfunction has been corrected. Each cutoff device should be tested independently of the other.

Quick Drain Test—The quick drain test is applicable to low-water fuel cutoff and alarm devices having actuating elements (floats or electrodes) located in a drainable chamber, external to the boiler shell. The test consists of blowing down the chamber at a time when the burner is operating. If the device is functional, it should cause the burner to shut off and the alarm to sound. It will also flush the chamber and connections of accumulated sediment. As with the slow drain test whenever dual cutoffs are installed, each device should be tested independently of the other. A quick drain test as described above is not applicable to so-called "built-in" low water cutoff or alarm devices having the actuating elements (float or electrodes) extending inside the boiler shell. **Note**: Such tests are not practical on operating hot water heating boilers, since the systems would have to be drained to perform them.

CENTRIFUGAL PUMPS

Commonly used to move liquids through piping, centrifugal pumps are one of the simplest machines made for and used in every type of facility. Their main components consist of a rotating impeller, a shaft, and a stationary casing (volute) and bearings. The impeller is used to create flow by the addition of energy to a fluid; the fluid entering the impeller along or near its rotating axis at the suction end is then accelerated (flowing radially outward into a diffuser or volute chamber (casing) where it exits into the downstream piping at the discharge end. See Figure I-10.

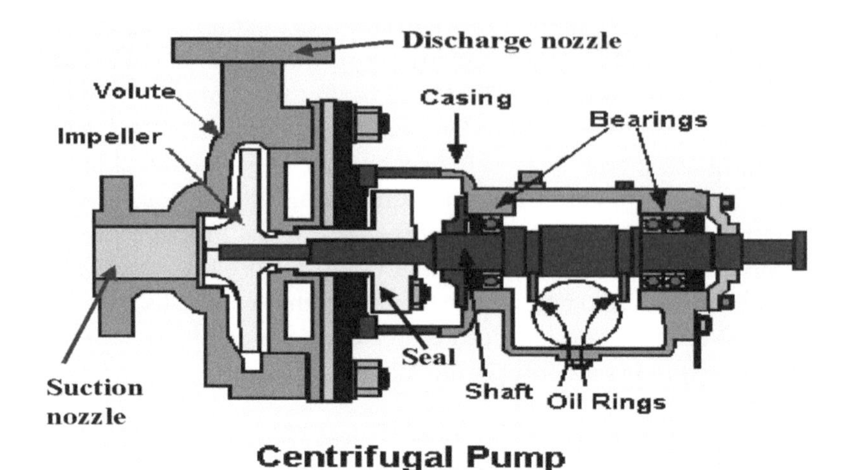

Figure I-10
Common Pump Parts

How They Work

Modern pumps have a spinning "impeller" which normally has backward-swept blades that directly push water outward. Most of the energy conversion is due to the outward force that curved impeller blades impart on the fluid. Some of the energy also pushes the fluid into a circular motion, and this circular motion can also convey some energy and increase the pressure at the outlet. See Figure I-11.

On the premise that they are of good design, properly installed, operated with due diligence and adequately maintained, these workhorses will provide years of completely trouble-free and satisfactory service. The main three problems encountered with these devices are poor design, improper operation and inadequate maintenance.

Operation

There are two basic requirements that must be met at all times for a trouble-free operation and long service life. The *first* is that no cavitation of the pump occurs throughout the broad operating range, and the *second* is that a minimum continuous flow is always maintained during operation. After starting the pump, periodically check the working condition of the pump, read the instruments such as gauges, amp

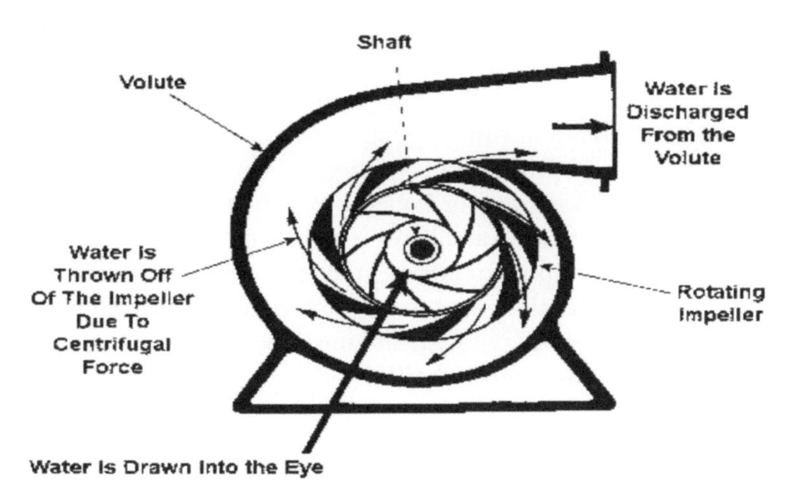

Figure I-11
Centrifugal Pump Impeller Action

meters, flow meters, etc., listen and feel for abnormal vibrations or noises. **Note**: if any of these problems is noticed, the pump should be stopped immediately, to search for the cause and make the necessary corrections. Check the pump/motor alignment, the running conditions of the bearings and the mechanical seals. If any deterioration of pump performance is experienced (not attributable to changes in system demands) the pump must be repaired or replaced.

Caveats

Centrifugal pumps should constantly perform as the installation requires. If at start-up there are suspicions of abnormal operation, it is recommended to stop the unit and investigate the causes. A clear understanding of the concept of cavitation, its symptoms, its causes and its consequences is essential in effective analyses and troubleshooting of cavitation problems.

CAUTION: NEVER OPERATE A PUMP DRY!

Common Maladies

There are a number of unfavorable conditions which may occur separately or simultaneously when a pump is operated at reduced flows, including excessive leakage from the casing, seal and stuffing box; deflection and shearing of shafts; seizure of pump internals; close tolerances erosion; separation cavitation; product quality degradation; excessive hydraulic thrust and premature bearing failures. Each condition may dictate a different minimum low flow requirement. The consequences of prolonged conditions of cavitation and low flow operation can be disastrous for both the pump and the process. Such situations must be avoided at all cost whether involving modifications in the pump and its piping or altering the operating conditions. Proper selection and sizing of the pump and its associated piping can not only eliminate the chances of cavitation and low flow operation but also significantly decrease their harmful effects.

CHILLERS

Compression Type

Compression type chillers are refrigeration systems that use refrigerant vapor-compression to produce chilled water which is used in a variety of applications, from space cooling to process uses. Chillers are the most commonly used refrigeration device for providing cooling for large structures. The main operating component within each chiller is the compressor, which condenses refrigerant gasses in the system, enabling the extraction of heat during the load cooling process. The compressor (powered by either an electric or internal combustion motor) increases the pressure on the gaseous refrigerant. The resulting hot, high-pressure gas is condensed to a liquid form by cooling it in a heat exchanger (condenser) that is exposed to an external environment (usually out-doors) where much of its heat is dissipated. The condensed refrigerant then passes through an orifice or a throttle valve into the load (evaporation) section. The lower pressure in the load section allows the liquid refrigerant to evaporate, which absorbs heat from the load. The resultant vaporized refrigerant then goes back into the compressor to repeat the cycle. Figure I-12 illustrates a typical plant installation, showing the chiller system cooling tower, condenser water and chilled water pump components and its cooling load application (shown as the air handling unit/cooling coil).

Chiller types (utilizing the compression process) differ from one another, based on the compressor technology they employ. They include reciprocating, rotary, centrifugal and scroll compressor types, as follows.

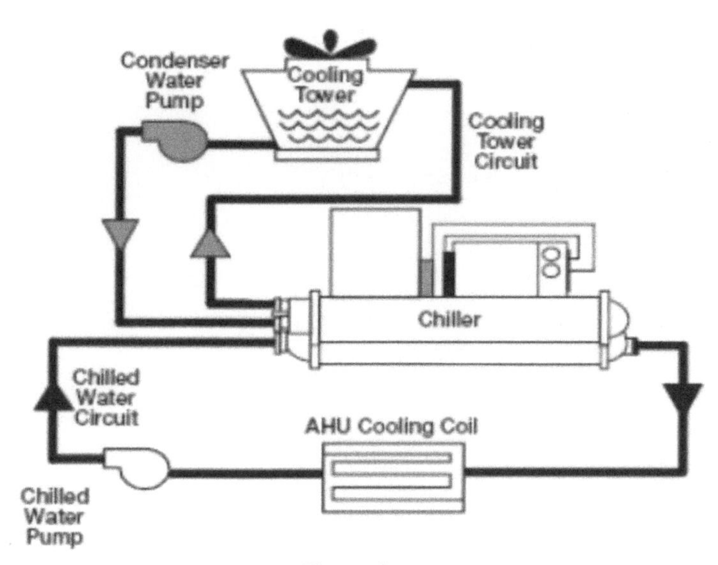

Figure I-12
Essential Chiller Components

Reciprocating Type

Reciprocating chillers utilize an internal piston contained within a cylindrical compartment. As gaseous refrigerant enters the compressor, the piston compresses the refrigerant to increase pressure. Once pressure levels have risen to a high enough point, an exhaust valve releases the compressed refrigerant so it can re-enter the cooling system and claim more heat energy from the load. See Figure I-13.

Rotary Type

Refrigerant gas in a rotary chiller is compressed using an internal roller that rotates within a steel cylinder. As refrigerant gas enters the cylinder through an intake valve, the rotating roller compresses it between the roller and the wall of the cylinder; then the rotating cylinder forces the compressed refrigerant out through an exhaust valve to complete the cooling cycle. See Figure I-14.

Centrifugal Type

In this type chiller, an impeller uses centrifugal force to accelerate refrigerant from an intake port to the walls of a cylinder, causing the refrigerant gas to collect along the walls, before being directed to an exhaust port to continue the cycle. See Figure I-15.

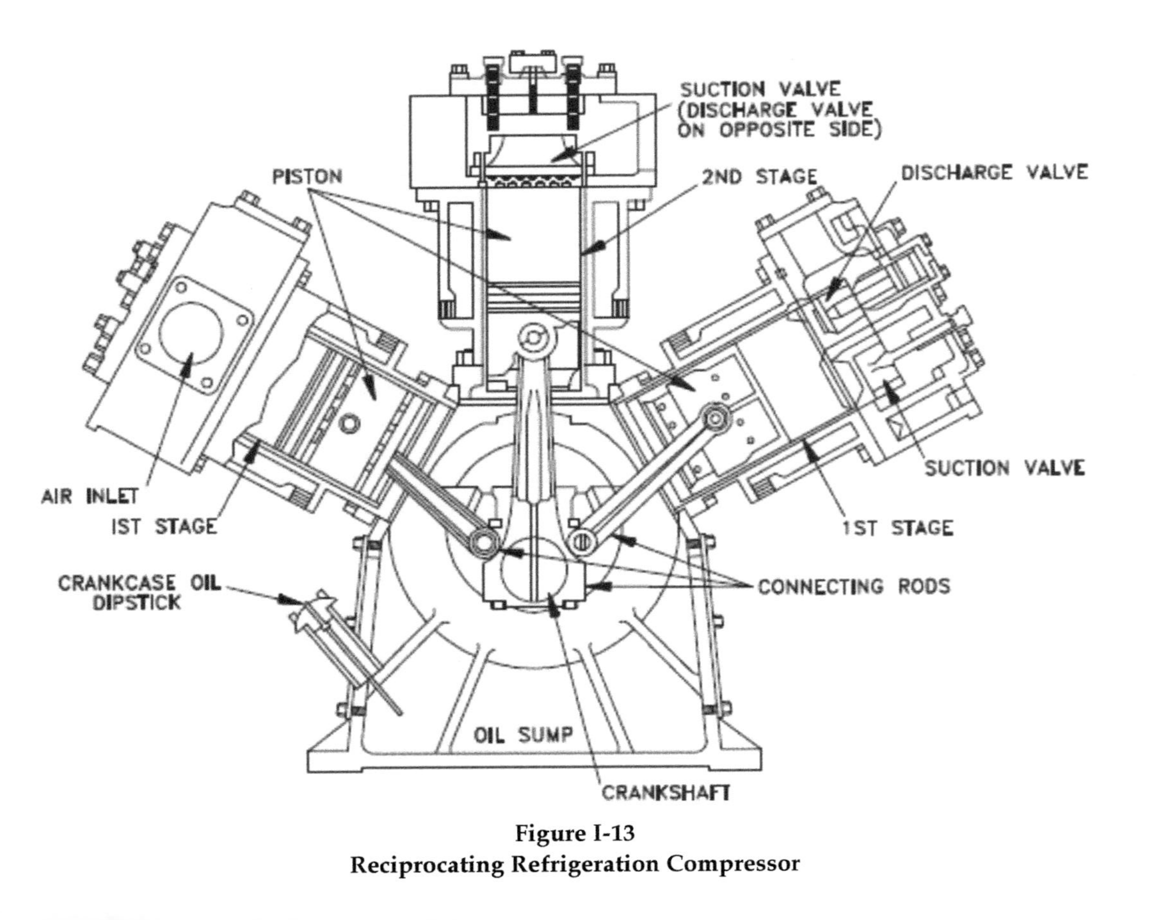

Figure I-13
Reciprocating Refrigeration Compressor

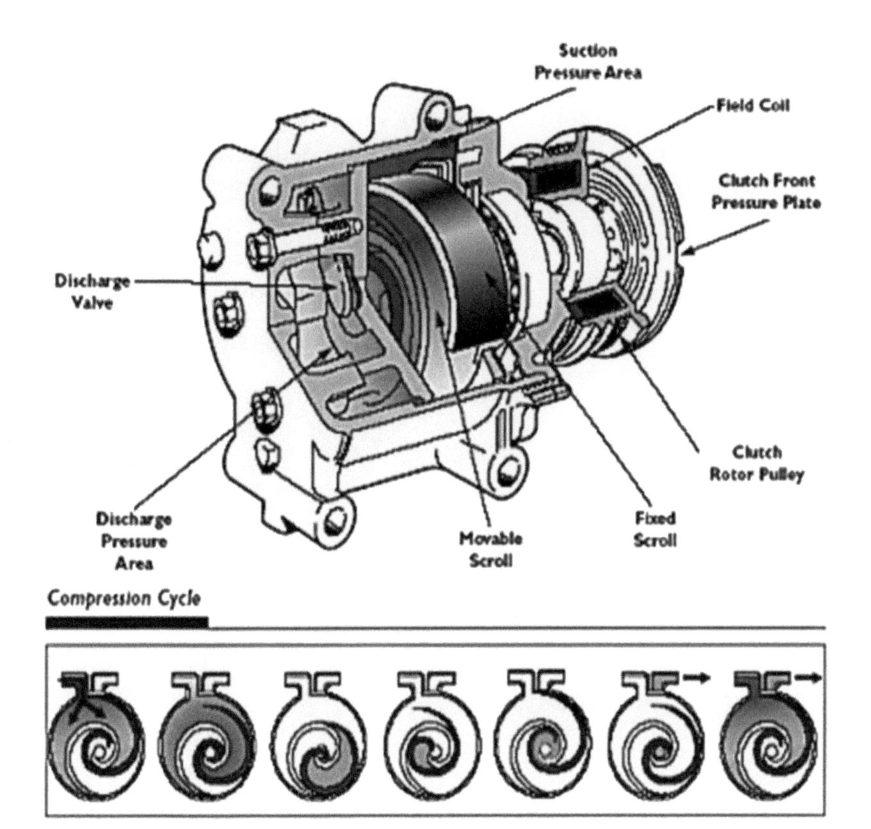

Figure I-14
Rotary Compressor Parts and Compression Cycle

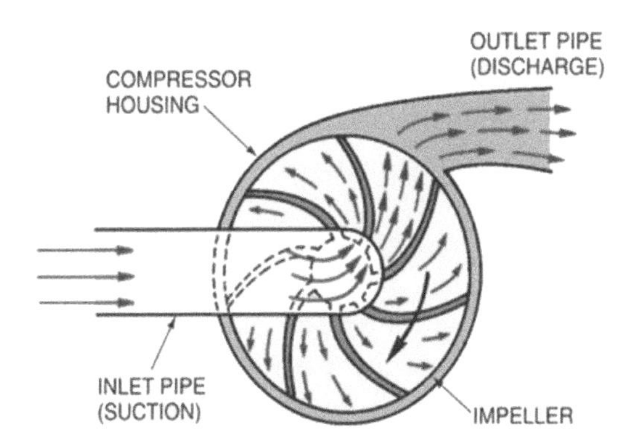

Figure I-15
Centrifugal Chiller Impeller Action

Scroll Type

Scroll compressors use two interlocking coils (one stationary and one rotating). Refrigerant gas gets caught between the walls of the coils when it enters the compressor and is compressed down into the center of the two coils, where it exits through an exhaust port. See Figure I-16.

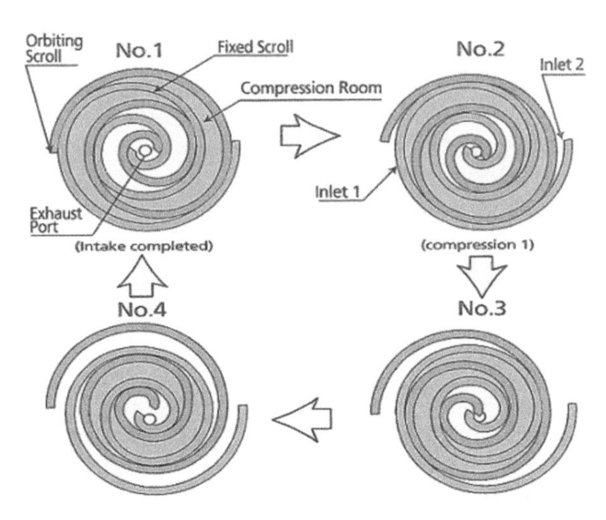

Figure I-16
Scroll Type Compression Action

Absorption Type

Absorption type chillers differ from compression type chillers in two distinct ways. First, in the type of refrigerant used (compression types are generally hydrocarbon-based, while absorption types utilize water or ammonia), then by the means with which they extract and dissipate heat. In lieu of compressing refrigerant vapor, absorption chillers dissolve the vapor in an absorbent and transfer the resulting product to a higher-pressure environment using a pump. The absorption cooling cycle employs three stages: evaporation, absorption and regeneration. In the evaporation stage, liquid refrigerant evaporates in a low (partial pressure) environment, where the temperature needed for evaporation is low. Here, heat is extracted from the load. In the absorption stage, the refrigerant gas is absorbed by another liquid such as brine (a salt solution). In the final (regeneration) stage, the now liquid refrigerant is heated, causing the refrigerant to evaporate. The hot refrigerant gas is then passed through a heat exchanger (condenser), transferring its heat outside the system, and then it condenses. The condensed (liquid) refrigerant is then returned to recommence the evaporation process. See Figure I-17.

CONDENSATE RETURN SYSTEMS

In a steam plant, condensate is the liquid phase produced by the condensation of steam within the delivery line and at the load device (as the work is accomplished). Condensate contains a high percentage of the energy (typically 16%) used to produce steam. Cost-effective energy conservation measures require condensate to be returned back to the boiler plant as hot as possible. System components should be designed and installed with maintenance in mind, utilizing connections that minimize leaks. Lacking a proactive maintenance plan will reduce the anticipated lifespan of the condensate system. The prime objective of the system components is that condensate must not be allowed to accumulate in the plant. Accumulated condensate will inhibit performance and encourage the corrosion of pipes, fittings and equipment. Condensate must not be allowed to accumulate in the steam main. Here it can be picked up by high velocity steam, leading to erosion and water hammer in the pipe work. See Figure I-18.

Inappropriate condensate connections will cause system problems. The primary issue will be water hammer that will result from the improper connection location. Flash steam introduced to the main condensate header due to an improper connection location will interact with cooler condensate causing water hammer. Water hammer is the leading cause of premature component failures in a steam/condensate system. It is imperative that all condensate branch lines are connected into the top dead center of the main condensate header on a horizontal plane. This cannot be overstated and there is no exception to this rule. Following are improper connections that can cause serious problems in your system.

- Connection to the bottom of a condensate header.
- Connection to the side of a condensate header.
- Connection to a vertical condensate header.

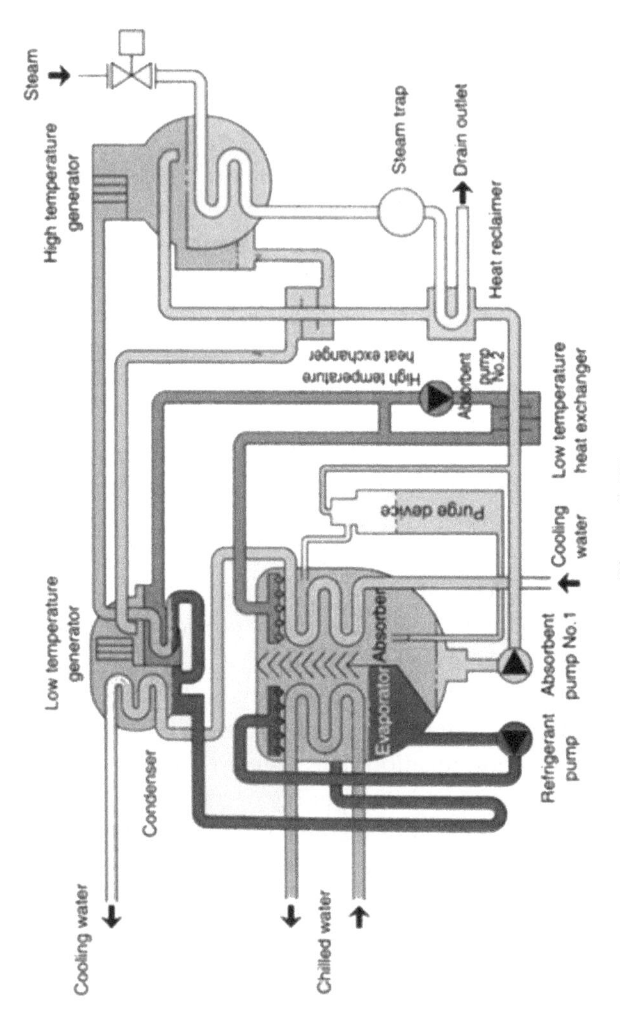

Figure I-17
Absorption Chiller

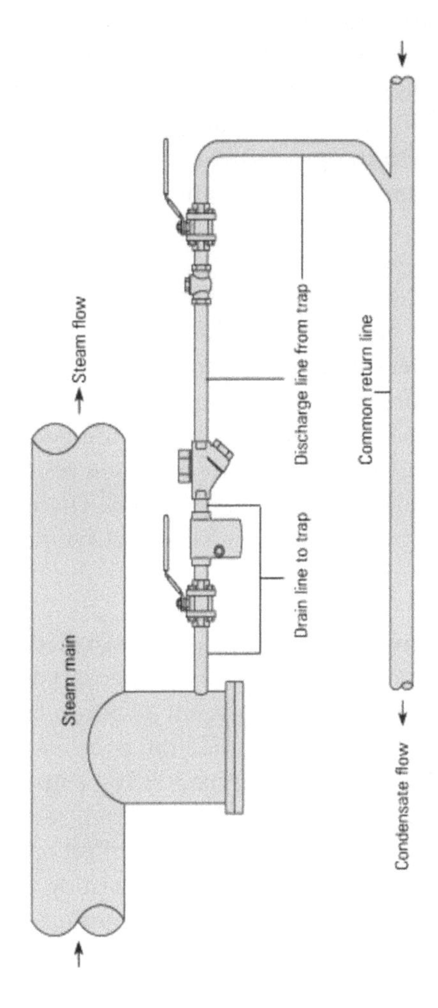

Figure I-18
Condensate Return on a Steam Main

The best method for improving steam system energy efficiency, reducing chemical costs, and reducing make-up water costs is to return the maximum quantity of condensate to the boiler plant. An important factor to increase overall steam system efficiency is to maximize the temperature of the returning condensate. This permits high thermal cycle efficiency for the overall steam system. Several factors that impact the reliability, performance, longevity, and maintenance requirements for the condensate piping system are condensate line sizing that factors condensate liquid, and flash steam quantities; location of the condensate line with respect to the process equipment; locations of the condensate branch line connection into the main condensate headers; and insulation techniques.

> **NOTE**: Although condensate is hot water, sizing a condensate line as if it were hot water would result in an undersized line. Undersized condensate lines will create excessive backpressure in the system and problems will occur in the system. The key item to remember is that there are two major differences between condensate and hot water. Condensate lines will contain two phases, condensate (liquid) and flash steam (vapor). Therefore, the correct size of a condensate line is somewhere between a hot water line and a steam line.

Condensate receivers (tanks) should be installed in a clean, dry, ventilated location which is accessible for inspection and care. The receiver inlet should be low enough to permit all return lines to empty by gravity to the receiver. No special foundation is necessary for the unit, although the floor or surface upon which it is to be installed should be structurally sound and relatively smooth and level. See Figure I-19.

Condensate pumps are devices used to collect and circulate condensate (water) back into a steam system for reheating and reuse. These pumps help reduce costs associated with the production of steam. To recover condensate, these pumps tend to run intermittently, switching on when a connected basin or tank has filled with condensate to a pre-determined level. They usually run intermittently and are attached to a tank (receiver) in which condensate can accumulate. As liquid (condensate) accumulates in the tank, it raises a float switch which energizes the pump. The pump runs until the level of liquid in the tank is substantially lowered. Some systems may include two pumps that operate

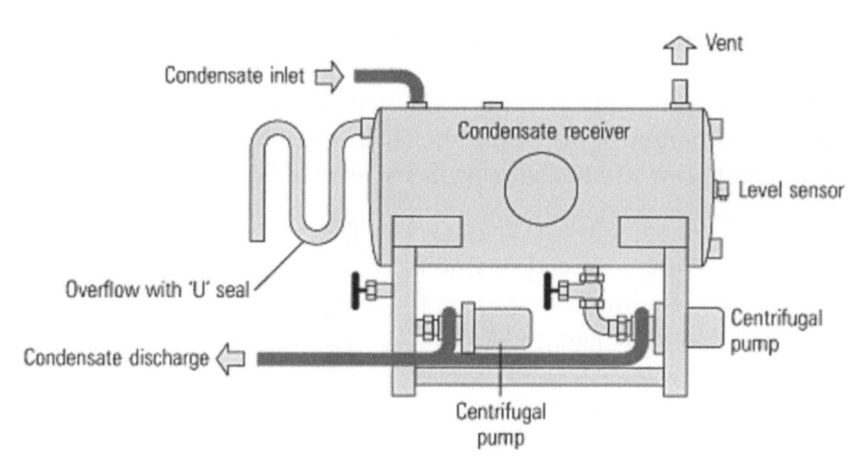

Figure I-19
Condensate Receiver Tank

alternately to service the tank and a two-stage switch which serves to energize the on-duty pump at the first stage (level) and the remaining pump at the second level. The second (backup) pump is provided in case the other pump fails. See Figure I-20.

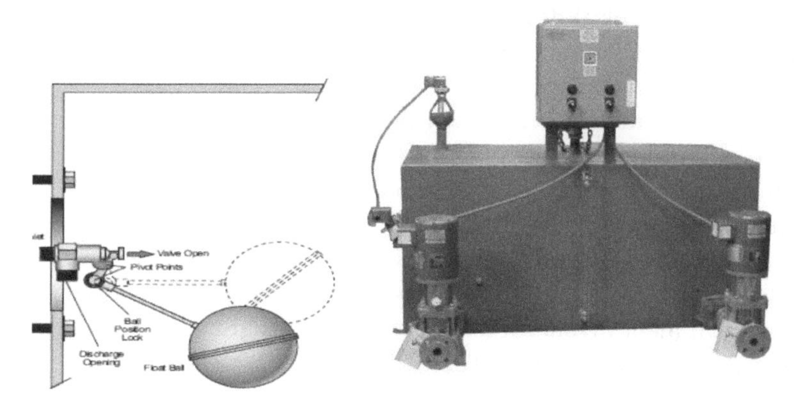

Figure I-20
Condensate Tank and Internal Float Mechanism

NOTE: Pressure gauges should be installed throughout the system. These inexpensive devices are a key aid in troubleshooting the steam and condensate system. Their presence will help to ensure a reliable and long life span of the condensate system.

Gauge glasses are transparent tubes through which an operator can observe the level of liquid contained within a tank or vessel. On condensate receivers, a glass tube is connected to the top of the tank at one end and the bottom of the tank at the other. The level of liquid observed in the sight glass will be the same as the level of liquid in the tank. See Figure I-21.

Note: Only properly trained personnel should install and maintain water gauge glass and connections. Wear safety glasses during installation. Before installing, make sure all parts are free of chips and debris. Keep gauge glass in original packaging until ready to install.

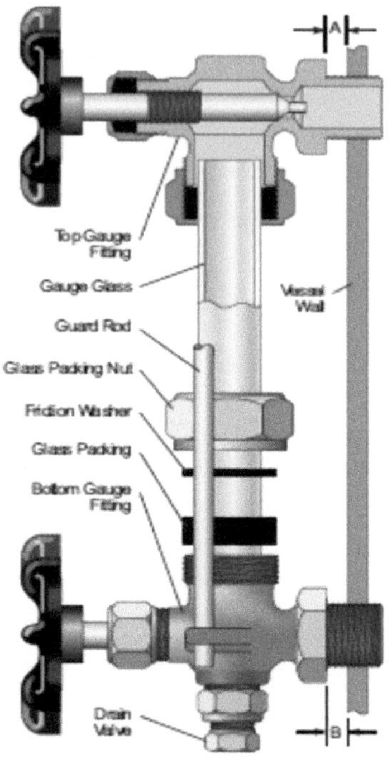

Figure I-21
Typical Receiver Tank Gauge Glass

COOLING TOWERS

To assure their continued operation, all mechanical devices require some degree of human intervention in their lifetimes. If there is an exception to that statement, it doesn't lie in the care needed by the oft ignored cooling tower. These major components in the physical plant scheme of operations are among the least understood, simplest to run and most neglected machines carried in our fixed equipment inventories. On cue from a programmer, they operate cyclically to reject the heat absorbed by the evaporators, chilled water loops and cooling coils which comprise our HVAC systems. Around the clock, season after season, they provide efficient, reliable service while battling the ravages of inclement weather, insufficient maintenance and poor water quality. Is it any wonder they sometimes just up and die?

Types of Towers

Cooling towers are manufactured in a variety of types, shapes and sizes. Classed as either cross-flow or counter flow for the flow relationship of the air and water within them, the particular design chosen for a given system will depend on many factors, including—spatial limitations, the availability of creek, lake or river waters for final rejection, the dependability of natural breezes and the calculated heat load configuration of the systems they serve. Regardless of their size, the arrangement of their internal parts or whether they are mounted on roofs or installed at ground level, cooling towers can be categorized as falling into one of three basic groups; either atmospheric, mechanical or hybrid.

Atmospheric Cooling Towers

Atmospheric cooling towers rely only on atmospheric conditions, using no mechanical devices to move air through the tower. Like natural draft chimneys, they rely on the natural movement of air relative to its density at different temperatures, to create flow through the tower. This natural induction is often aided by the introduction of water under pressure, in the form of a spray. These towers are easily affected by adverse wind conditions and are of limited use in applications calling for consistent cold water temperatures. The hyperbolic towers located at electrical power generation plants are examples of natural draft type units. See Figure I-22.

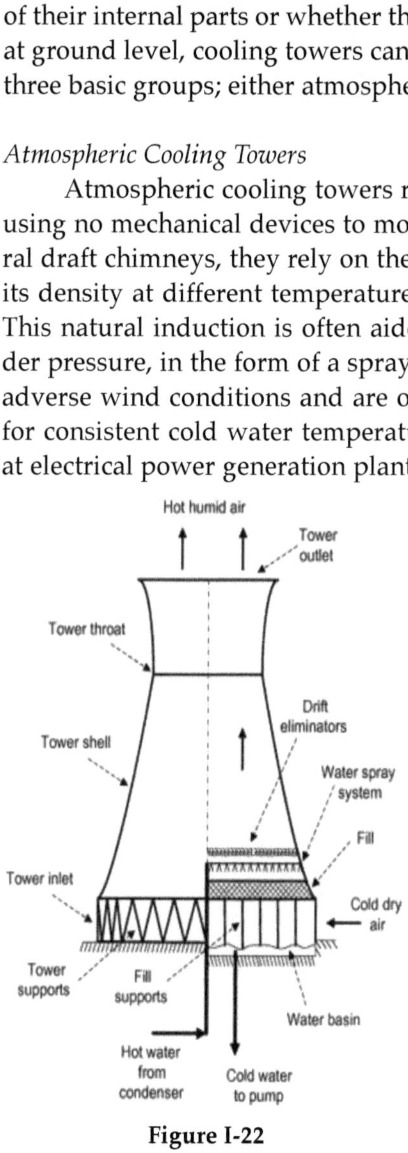

Figure I-22
Natural Draft Tower Type

Mechanical Draft Towers

Mechanical draft towers (like mechanical draft chimneys) utilize forced and induced draft fans to provide predetermined air volumes through the towers in which they are mounted. They overcome the adversities associated with random wind, suffered by natural draft towers. In these units, air flow can be regulated to compensate for changing atmospheric and load conditions, providing greater stability of thermal performance in the face of psychrometric variables. See Figure I-23.

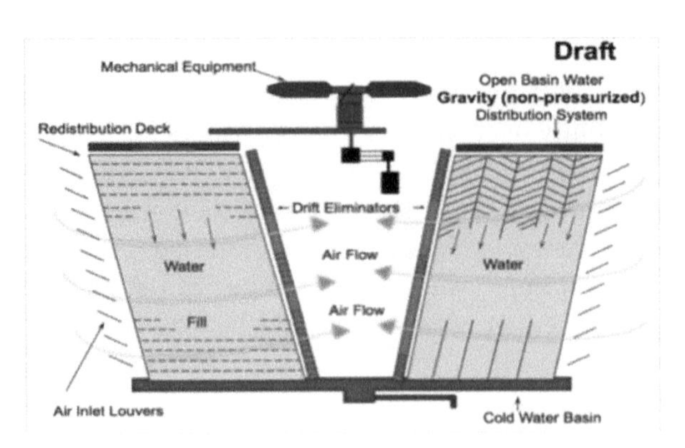

Figure I-23
Air and Water Flow in a Cooling Tower

Hybrid Cooling Towers

Hybrid cooling towers are fan-assisted, natural draft units designed to minimize the cost of stack construction and the horsepower required to provide air movement through them. The fans are designed to operate during peak load and to elevate the level of heat discharged from the tower. See Figure I-24.

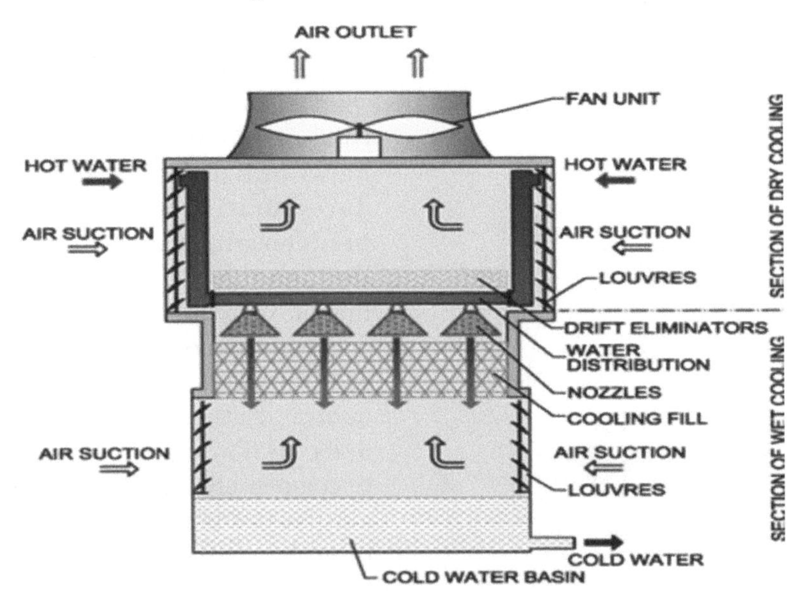

Figure I-24
Internal View of a Hybrid Cooling Tower

System Design and Performance

Opt for proven reliability, reasonable expandability, maximum flexibility, and ultimate safety and installation as close to the load as possible. Take into account: performance, proper setting/orientation, sizing for the load being handled, proper clearances/tolerances, reliable operating/safety controls, the ability to meter loading, proper instrumentation, coordinated protective devices, access for maintenance, usage/adaptability and available spare parts. No device or system has ever been designed or installed that doesn't have the potential to malfunction. Time alone predisposes them to failure. Check on issues such as insufficient air supplies and improper water quality, either of which may cause your towers, ergo your chillers, to cease functioning. Inadequate air supply volumes can result from the installation of walls, wind-breaks, privacy fences, shrubbery located near units for aesthetic purposes, by blockage of air intake louvers, or for a variety of reasons, not the least of which are autumn leaves. Air circulation problems can also be experienced due to down drafts and the discharging of exhausts and hot gasses from vents and stacks in close proximity to the tower's air passages. Broken or deteriorated internal components can also wreak havoc with air and/or water circulation. Of all the things that can and do go wrong in cooling towers, none is more contributory to failure than poor quality water, which finds its way into the tower as the direct result of an inadequate treatment program, to choke off passages, corrode system structures and breed disease-bearing pathogens such as the Legionnaires' bacillus.

ELECTRIC MOTORS

Electric motors are extremely functional, diverse electromechanical devices that convert electrical energy into mechanical energy, through the interaction of magnetic fields and current-carrying conductors. The reverse process, producing electrical energy from mechanical energy, is accomplished by a generator or dynamo. (Electric motors can be run as generators and vice versa.) These machines range in size from "small" (like those used to operate electric wristwatches), "medium" (typical of those found in facility physical plants), and "large" (used to transport pipeline oil and propel locomotives). Electric motors may be classified by the source of electric power, by their internal construction and by their application.

Categories

The main distinctions between motors (as shown in Figure I-25) are:

- Whether they are DC (direct current driven) motors or AC (alternating current driven) motors.
- Whether they are "synchronous" (in which the rotation rate of the shaft is synchronized with the frequency of the AC supply) or induction type "asynchronous," requiring SLIP (relative movement between the magnetic field [generated by the stator] and a winding set [the rotor] to induce current in the rotor to generate torque).
- Whether they are "rated horsepower" (output power) or "fractional horsepower." Motors of less than 746 watts are often referred to as fractional horsepower motors (FHP).
- Whether they are "single phase" (a type of motor with low horsepower that operates on 120 or 240 volts) or "polyphase" (a three-phase motor with a continuous series of three overlapping AC cycles offset by 120 degrees).
- Whether they are "capacitor start" (a capacitor is in series with a starting winding and provides more than double the starting torque with one third less starting current than the split phase motor, for loads which are hard to start). The capacitor and starting windings are disconnected from the circuit by an automatic switch when the motor reaches about 75% of its rated full-load speed. Or whether they are "split phase" (a single-phase motor that consists of a running winding, starting winding, and centrifugal switch which has start and run windings, both of which are energized when the motor is started). When the motor reaches about 75% of its rated full-load speed, the starting winding is disconnected by an automatic switch.

Maintenance Considerations

Scheduled routine inspection and service are critical for ensuring the operating integrity of your plant's electrical motors. Depending on the service environment, operation and application to which they are applied, the frequency of service calls can vary widely. A motor may require additional or more frequent attention if a breakdown would cause health or safety problems, severe loss of production, damage to

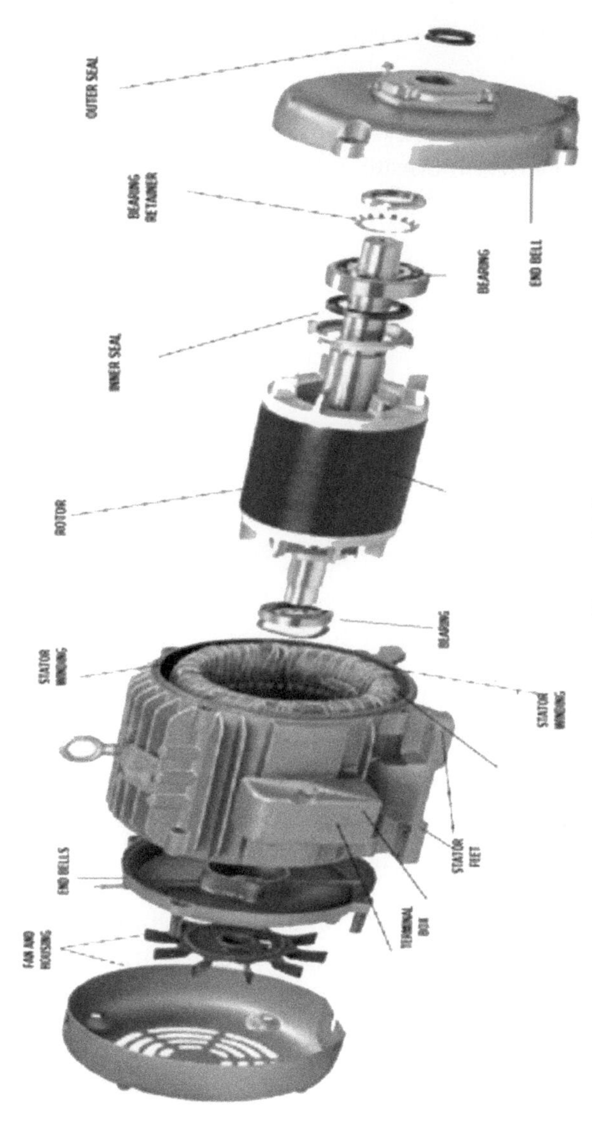

Figure I-25
AC (3 Phase) Motor Components

expensive equipment or other serious losses. But, as a general rule, including the motors in the maintenance schedule for the driven machine or general plant equipment is usually sufficient. Routine inspection and servicing can usually be done without disconnecting or disassembling the motor and involves the following factors:

Lubrication

Lubricate the bearings only when scheduled or if they are noisy or running hot. Do *not* over-lubricate. Excessive grease and oil creates dirt and can damage bearings. See Figure I-26.

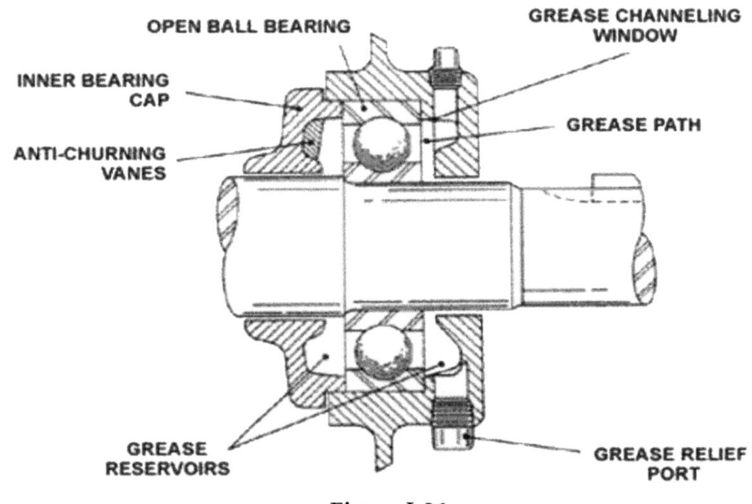

Figure I-26
Ball Bearing Grease Path

Winding Insulation

When records indicate a tendency toward periodic winding failures in the application, check the condition of the insulation with an insulation resistance test. Such testing is especially important for motors operated in wet or corrosive atmospheres or in high ambient temperatures.

Heat, Noise and Vibration

Feel the motor frame and bearings for excessive heat or vibration. Listen for abnormal noise. All indicate a possible system failure. Promptly identify and eliminate the source of the heat, noise or vibration.

Dirt and Corrosion

Wipe, brush, vacuum or blow accumulated dirt from the frame and air passages of the motor. Dirty motors run hot when thick dirt insulates the frame and clogged passages reduce cooling air flow. Heat reduces insulation life and eventually causes motor failure. Feel for air being discharged from the cooling air ports. If the flow is weak or unsteady, internal air passages are probably clogged. Remove the motor from service and clean. Check for signs of corrosion. Serious corrosion may indicate internal deterioration and/or a need for external repainting. Schedule the removal of the motor from service for complete inspection and possible rebuilding. In wet or corrosive environments, open the conduit box and check for deteriorating insulation or corroded terminals. Repair as needed.

Brushes and Commutators (DC Motors)

Observe the brushes while the motor is running. The brushes must ride on the commutator smoothly with little or no sparking. Stop the motor. Be certain that:

- The brushes move freely in the holder, and the spring tension on each brush is about equal.
- Every brush has a polished surface over the entire working face, indicating good seating.
- The commutator is clean, smooth and has a polished brown surface where the brushes ride.
- There is no grooving of the commutator (small grooves around the circumference of the commutator).

Replace the brushes if there is any chance they will not last until the next inspection date. If accumulating, clean foreign material from the grooves between the commutator bars and from the brush holders and posts. Brush sparking, chatter, excessive wear or chipping, and a dirty or rough commutator indicate motor problems requiring prompt service.

Motor Duty

Light Duty

Motors operate infrequently (1 hour/day or less) as in power tools, valves, door openers.

Standard Duty

Motors operate in normal applications (1 or 2 work shifts). Examples include air conditioning units, conveyors, refrigeration apparatus, laundry machinery, water pumps, machine tools, garage compressors.

Heavy Duty

Motors are subjected to above normal operation and vibration (running 24 hours/day, 365 days/year). Such as in motor-generator sets, fans, pumps.

Severe Duty

Extremely harsh, dirty motor applications. Severe vibration and high ambient conditions often exist.

Excessive Friction

Misalignment, poor bearings and other problems in the driven machine, power transmission system or motor increase the torque required to drive the loads, raising motor operating temperature.

Electrical Overloads

An electrical failure of a winding or connection in the motor can cause other windings or the entire motor to overheat.

Windings

Acceptable motor life requires selection of the proper enclosure to protect the windings from excessive dirt, abrasives, moisture, oil and chemicals. Such motors can be removed from service and repaired before unexpected failures stop production. Whenever a motor is opened for repair, service the windings as follows:

- Accumulated dirt prevents proper cooling and may absorb moisture and other contaminants that damage the insulation. Vacuum the dirt from the windings and internal air passages. Do not use high pressure air because this can damage windings by driving the dirt into the insulation.
- Abrasive dust drawn through the motor can abrade coil noses, removing insulation. If such abrasion is found, the winding should be re-varnished or replaced.

- Moisture reduces the dielectric strength of insulation which results in shorts. If the inside of the motor is damp, dry the motor per information in cleaning and drying windings.
- Wipe any oil and grease from inside the motor. Use care with solvents that can attack the insulation. If the insulation appears brittle, overheated or cracked, the motor should be re-varnished or, with severe conditions, rewound.
- Loose coils and leads can move with changing magnetic fields or vibration, causing the insulation to wear, crack or fray. Re-varnishing and retying leads may correct minor problems. If the loose coil situation is severe, the motor must be rewound.
- Check the lead-to-coil connections for signs of overheating or corrosion. These connections are often exposed on large motors but taped on small motors. Repair as needed.
- Check wound rotor windings as described for stator windings. Because rotor windings must withstand centrifugal forces, tightness is even more important.
- In addition, check for loose pole pieces or other loose parts that create imbalance problems.

ELECTRICAL DISTRIBUTION SYSTEMS

Power plants in the United States produce electricity, using fuel sources such as coal, oil, natural gas, nuclear energy, hydropower and alternate fuels (geothermal energy, wind power, biomass, and solar energy), which is loaded onto the nation's electrical grid. From the grid, electric power is drawn for diverse consumption throughout the country. Electrical systems are designed to supply customers with safe, reliable and affordable energy, requiring the application and incorporation of myriad complex processes and systems. Not all systems are designed exactly alike because each customer has their own special needs and geography; however, the same basic components are employed (as described below). See Figure I-27.

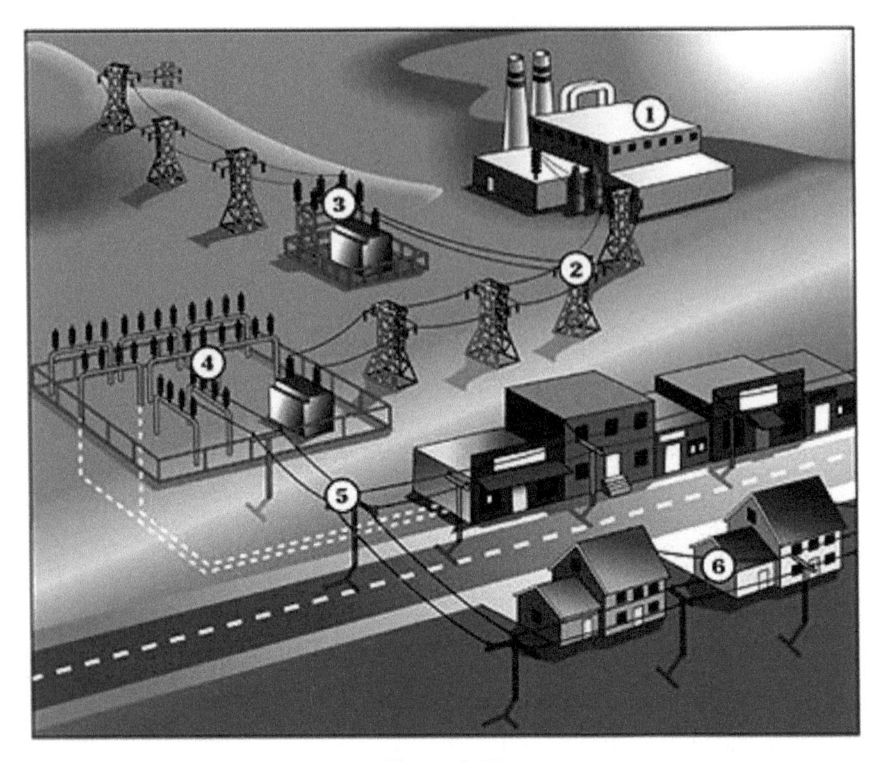

Figure I-27
Typical electric utility distribution

Power Plant

Electricity starts here, produced by spinning generators that are driven by water, a diesel engine, or a natural gas or steam turbine. Steam is made by burning coal, oil or natural gas or by a nuclear reactor. When needed, extra power is brought into an electric system from outside the plants.

Power Grid

Electricity is carried over a network, or grid, that connects power plants to a substation and from there to distribution lines that take the power to homes or businesses.

Transmission Substation

These facilities connect to wires emanating from the power plant. Here, large transformers increase voltage from thousands to hundreds of thousands of volts, enabling the power to be transmitted over long distances.

Distribution Substation

They are those small fenced-in areas that have electric lines coming in and going out. Inside these fenced-in areas are transformers that reduce voltage to a lower level so the power can be sent out on distribution lines to the surrounding community.

Distribution System

Includes main or primary lines and lower voltage or secondary lines that deliver electricity through overhead or underground wires to homes and businesses.

Service Connection

The line that connects to the meter. The meter is used to determine how many kilowatt-hours are used by each customer.

Plant Distribution

At the point of entry, the "in plant" (internal) distribution of electricity, beginning at the service connection at the building entrance, typically has distribution values ranging from 110 volts to 35 kV. Those values are further distinguished as that for primary distribution (voltages between 2.4 and 35 kV) and subsequently for secondary distribution (110 to 600 volt) systems. The goal of an electrical distribution system is the economical and safe delivery of adequate electric power for lighting, process and operation of electrical equipment.

Overhead Service Wiring

The electrical circuit between the utility company's mains and the customer's wiring is called the service. Service connections are sets of wires that are tapped onto the secondary mains and connected to the customer's wiring. These wires are also known as the service drop. An overhead service consists of wires and cables extending from a pole carrying the main to a point on the customer's building. See Figure I-28.

Underground Service Wiring

Underground services consist of plastic- or lead-covered cables which extend from the customer's service point to the mains to which they are connected. Like overhead services, they may consist of two or more conductors, which may be in the form of two or more single

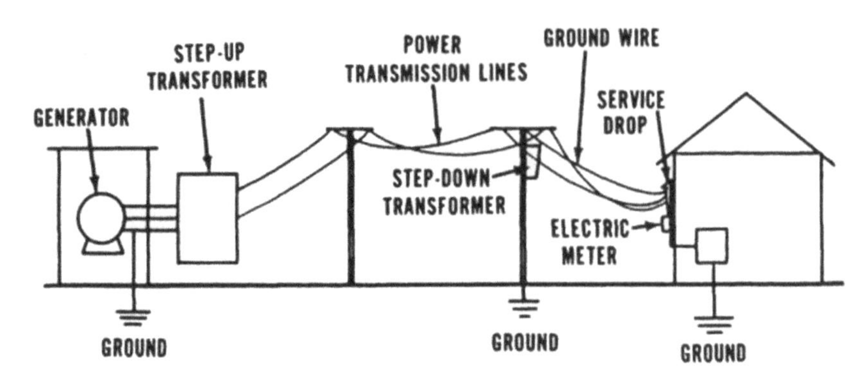

Figure I-28
Overhead Electrical Lines

conductor cables or one multi-conductor cable. These service cables are often installed in a steel or concrete duct, for at least the portion of the customer's premises, for safety reasons.

Voltage Phasing

A voltage phase is the difference of electrical potential between two points in an electrical circuit (the measurement and force of electricity that is being transmitted to a specific destination). Transformers regulate the amounts of electrical input transmitted to power sources from utility lines to meet power demands. When utilities send electricity through power lines to a residence or business, voltage phases are the driving force behind the electrical current. Not all equipment and machinery is designed to operate on every measure of voltage phase. It is necessary that voltage phases are controlled, so that load demands for equipment and machinery are met, otherwise damage could occur in your equipment.

Single-phase Systems

Single-phase systems use three wires (two hotwires and one neutral wire). Use of the three wires in different electrical combinations can provide different voltages. For example, one circuit can be made up of one wire and the neutral wire. That circuit provides 120 volts AC (alternating current)—the voltage required for most household lighting and small appliances. Another circuit can be made up of both hot wires to provide 240 volts AC, the voltage required for larger appliances. See Figure I-29.

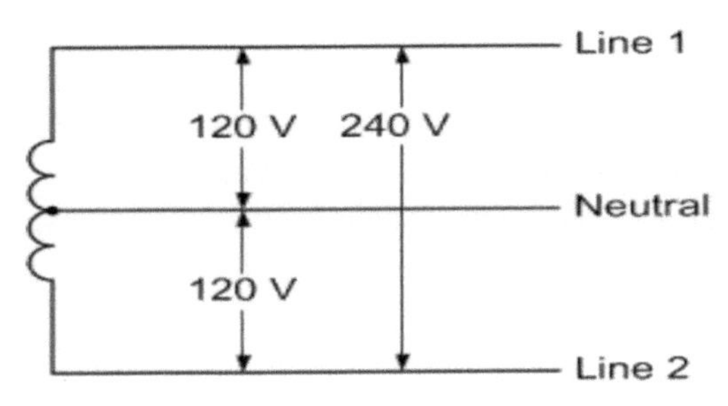

Figure I-29
Single-phase electrical schematic

Three-phase Systems

Three-phase systems are efficient and make equipment run more smoothly than single-phase systems. The reason is because each phase is 120 degrees apart; therefore, the equipment doesn't see a zero point. This is especially important when running motors because each new phase keeps the motor turning. As they rotate through the magnetic field, they generate power. Each phase is 120 degrees (out of phase) from each other. Each phase flows from the generator in a separate cable. The phases are delivered to the end users as either a three-phase or a single-phase power supply. (Single-phase power is available from a three-phase power system by using only one or two of the phases). This is beneficial for supplying power for lights, receptacles, heating and air-conditioning. There are two types of three-phase systems: three-phase/three-wire and three-phase/four-wire. See Figure I-30.

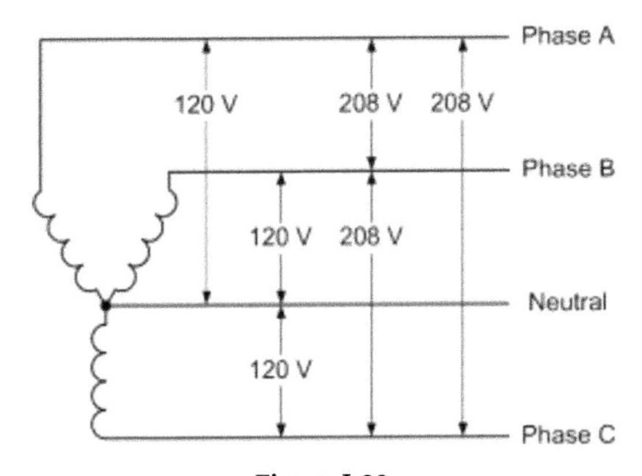

Figure I-30
Multi-phase electrical schematic

System Components

Distribution systems used in commercial and industrial locations are complex. They consist of metering devices to measure power consumption, main and branch disconnects, protective devices, switching devices (to start/stop power flow), conductors, and transformers. Power may be distributed to and through several switchboards, transformers and panel boards. See Figure I-31.

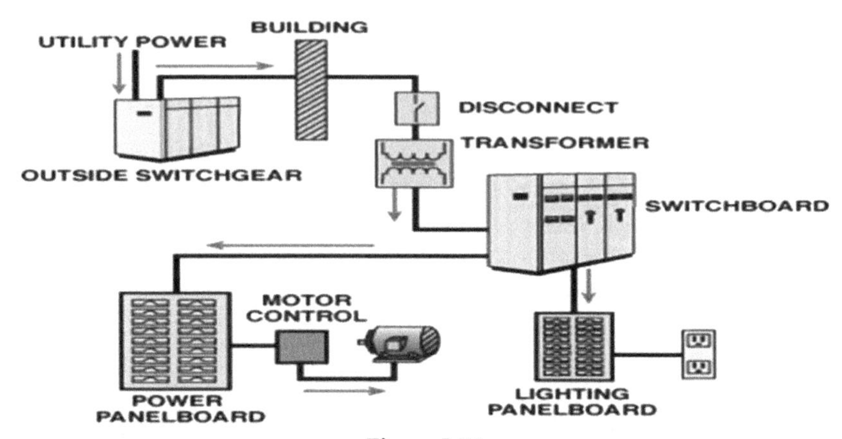

Figure I-31
Distribution of Electrical Power in a Building

Electrical Grounding

There are three important distinctions that must be understood regarding the function of electrical grounding at a building:

- *Grounding to trip the circuit breaker or blow the fuse* is necessary to protect people from electric shock by providing a good electrical path to route a faulty electrical connection (such as a short circuit) to ground (literally, to the earth) so that current will flow through and thus trip the circuit breaker or blow the fuse, safely and quickly turning off the electrical circuit. (A short circuit is one of the two ways that a fuse or circuit breaker will blow. The second is the drawing of more current [amps] than the circuit is intended to carry.)

- *Grounding to route stray electricity* can protect the insulation on electrical wires from damage due to high voltage from static electricity, electrical power surges, lightening, etc. to ground. The protection against static electricity helps reduce the chances of an unwanted spark that can cause damage to electronic components or cause an explosion of nearby combustible gases.

- *Grounding to provide a normal path for electricity to flow* through the electrical panel, though building circuits (and electrical devices connected to them) to the ground, to earth. If there is no good connection to earth, electricity does not flow. Ground and neutral are related electrical terms. Neutral is used to describe the "return path" from an electrical circuit to the electrical panel. Inside the main electrical panel, the neutral wires are bonded to the ground wires and to a grounding conductor that connects that assembly to earth, typically through a grounding electrode or "ground rod."

System Protection

The primary goal of all electrical power distribution systems is to provide power to electrical equipment with the utmost safety. Over current is a current that is higher than the amount of current a conductor or piece of equipment can carry safely. An overcurrent condition left unchecked can cause insulation and/or equipment damage as a result of excessive temperature and/or dynamic stresses. There are three types of overcurrent conditions: overloads, short circuits and ground faults.

Overloads are the result of placing excessive loads on a circuit, beyond the level the circuit was designed to handle safely. Insulation deterioration in electrical conductors is most often the result of such overload conditions. When an overload condition exists, the temperature of the conductor increases and, if left unchecked, could damage or ignite the conductor's insulation.

Short circuits (frequently called faults) are usually caused by abnormally high currents that flow when insulation on a conductor fails. When the insulation that protects one phase from another—or one phase from the ground—breaks down, hazardous short circuit currents can be expected to flow to ground.

Note: The short circuit condition must be eliminated quickly to prevent a fire and protect against damage to the system.

Ground faults are a particular type of short circuit. They are short circuits between one of the phases and ground. It is probably the most common low level fault experienced, especially on lower voltage circuits. Ground fault currents are often not large in magnitude and can go undetected for a period of time.

> **Note**: This type of fault might occur in the electrical outlets located in areas where water could be present.

HVAC SYSTEMS

The term HVAC (heating, ventilation and air-conditioning) system is used to refer to the equipment that can provide heating, cooling, filtered outdoor air, and humidity control to maintain comfort conditions in a building. Not all HVAC systems are designed to accomplish all of these functions. Some lack mechanical cooling equipment (AC), and many function with little or no humidity control. HVAC systems range in complexity from stand-alone units that serve individual rooms to large, centrally controlled systems serving multiple zones in a building.

In large buildings with heat gains from lighting, people, and equipment, interior spaces often require year-round cooling. Rooms at the perimeter of the same building (i.e., rooms with exterior walls, floors, or roof surfaces) may need to be heated and/or cooled as hourly or daily outdoor weather conditions change. In buildings over one story in height, perimeter areas at the lower levels also tend to experience the greatest uncontrolled air infiltration.

The features of the HVAC system in a given building will depend on several variables, including age of the design, climate and current weather conditions, building codes at the time of the design, budget that was available for the project, planned use of the building, owners' and designers' individual preferences and subsequent modifications. HVAC systems require preventive maintenance and prompt repairs if they are to operate correctly and provide comfortable conditions. Creature comfort is contingent on acceptable temperature and humidity ranges. In addition to thermal comfort, the control of relative humidity is important to limit the growth of microorganisms such as mold and dust mites. To control microorganisms, it is best to keep relative humidity below 60% (to control mold) and 50% (to control dust mites) at all times, including unoccupied hours. High relative humidity can foster proliferation of mold and dust mites. Table I-1 shows typical air quality values.

Table I-2
Typical Air Quality Values

Measurement Type	Winter	Summer
Dry Bulb at 30% RH	68.5°F - 76.0°F	74.0°F - 80.0°F
Dry Bulb at 50% RH	68.5°F - 74.5°F	73.0°F - 79.0°F
Wet bulb maximum	64°F	68°F
Relative humidity *	30% - 60%	30% - 60%

Basic Components

Some basic components of an HVAC system that deliver conditioned air to maintain thermal comfort and indoor air quality are: outdoor air intakes, mixed-air plenum and outdoor air control, air filters and filter media, heating and cooling coils, humidification and/or de-humidification equipment, supply fans, ducts, terminal devices, return air systems, exhaust or relief fans and air outlets, controls, self-contained heating or cooling units, boilers, cooling towers and water chillers. Figure I-32 shows the general relationship between many of these components; however, many variations are possible.

Note: Working with the electrical components of an HVAC system involves the risk of electrocution and fire. A knowledgeable member of the building staff should oversee the inspection of the HVAC controls functionality.

Coils and Drain Pans

Malfunctioning coils, including dirty coils, can waste energy and cause thermal discomfort. Leaky valves that allow hot or chilled water

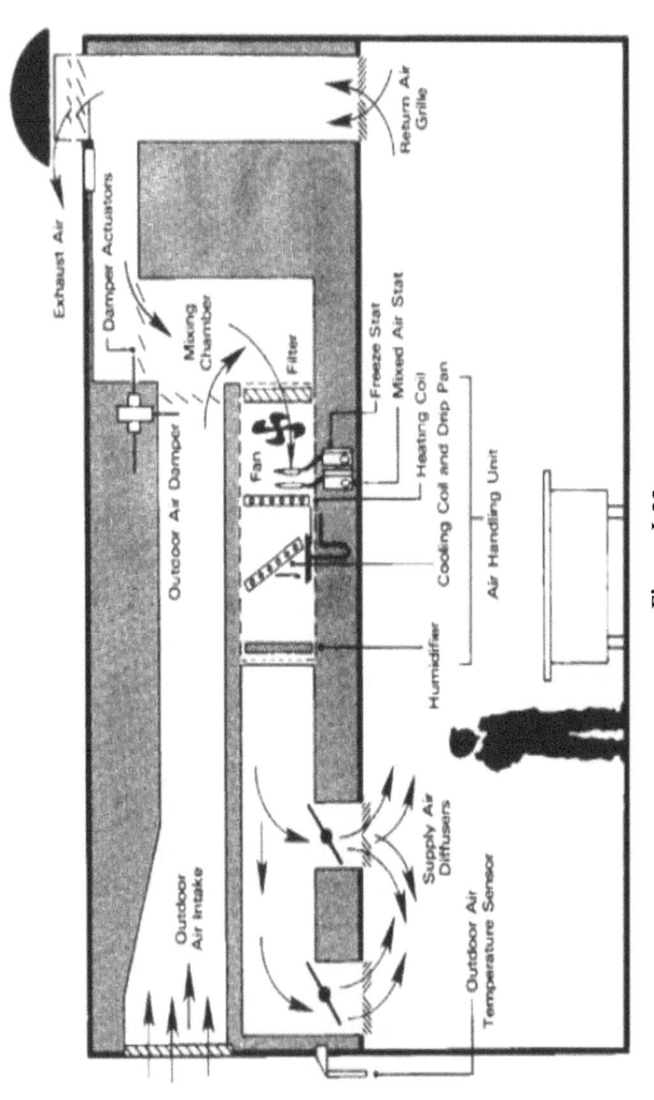

Figure I-32
Internal View of Air Handling Unit Components

through the coil (when there is no demand) waste energy and create thermal discomfort. Cooling coils dehumidify the air and cause condensate water to drip into a drain pan and exit via a deep seal trap. Standing water will accumulate if the drain pan is not properly designed and maintained, creating a microbial habitat. Proper sloping and frequent cleaning of the drain pans is essential to good indoor air quality.

Humidification and Dehumidification Equipment

Potable water rather than boiler water should be used as a source of steam to avoid contaminating the indoor air with boiler treatment chemicals. Wet surfaces should be properly drained and periodically treated as necessary to prevent microbial growth. Duct linings should not be allowed to become moist from water spray.

Air Filters

Use filters to remove particles from the air stream. Filters should be replaced on a regular basis, on the basis of pressure drop across the filter, or on a scheduled basis. Fans should be shut off when changing the filter, to prevent contamination of the air. Filters should fit tightly in the filter housing. Low efficiency filters (ASHRAE Dust Spot rating of 10%-20%), if loaded to excess, will become deformed and even "blow out," leading to clogged coils, dirty ducts, reduced indoor air quality and greater energy use. Higher efficiency filters are often recommended as a cost-effective means of improving IAQ performance while minimizing energy consumption. Filtration efficiency should be matched to equipment capabilities and expected airflows.

Outdoor Air Dampers

Screens and grilles can become obstructed. Remove obstructions, check connections, and otherwise insure that dampers are operating to bring in sufficient outdoor air to meet design-level requirements under all operating conditions.

Ducts

Duct leakage can cause or exacerbate air quality problems and waste energy. Sealed duct systems with a leakage rate of less than 3% will usually have a superior life cycle cost analysis and reduce problems associated with leaky ductwork. Common problems include leaks around loose-fitting joints and leaks in return ducts in unconditioned spaces or underground which can draw contaminants from these spaces into the supply air system. A small amount of dust on duct surfaces is normal. Parts of the duct susceptible to contamination include areas with restricted airflow, duct lining, or areas of moisture or condensation. Problems with biological pollutants can be prevented by: minimizing dust and dirt build-up (especially during construction), promptly repairing leaks and water damage, keeping system components dry that should be dry, cleaning components such as coils and drip pans, good filter maintenance and good housekeeping in occupied spaces.

Exhaust Systems

In general, slightly more outdoor air should be brought into the building than the exhaust air and relief air of the HVAC system. This will insure that the building remains under slight positive pressure. Exhaust intake should be located as close to the source as possible. Fans should draw sufficient air to keep the room in which the exhaust is located under negative pressure relative to the surrounding spaces, including wall cavities and plenums. Air should flow into, but not out of, the exhaust area, which may require louvered panels in doors or walls to provide an unobstructed pathway for replacement air. The integrity of walls and ceilings of rooms to be exhausted must be well maintained to prevent contaminated air from escaping into the return air plenum. Provisions must be made for replacing all air exhausted out of the building with make-up outside air.

Return Air Plenum

Space above the ceiling tiles is often used as a return air plenum. Strictly follow code which restricts material and supplies in the plenum to prevent contamination and insure that airflow is not interrupted. Remove all dirt and debris from construction activity. All exhaust systems passing through the plenum must be rigorously maintained to prevent leaks, and no exhaust should be released into the plenum. Avoid condensation on pipes in plenum area. Moisture creates a habitat for microbial growth.

System Types

Note: Refer to areas regulated by a common control (single thermostat) as zones.

Single Zone

A single air handling unit can only serve more than one building area if the areas served have similar heating, cooling, and ventilation requirements, or if the control system compensates for differences in heating, cooling, and ventilation needs among the spaces served.

Multi-Zone Systems

These systems can provide each zone with air at a different temperature by heating or cooling the airstream in each zone.

Constant Volume Systems

Deliver a constant airflow to each space. Changes in space temperatures are made by heating or cooling the air or switching the air handling unit on and off, not by modulating the volume of air supplied. These systems often operate with a fixed minimum percentage of outdoor air or with an "air economizer" feature.

Variable Air Volume Systems

Maintain thermal comfort by varying the amount of heated or cooled air delivered to each space, rather than by changing the air temperature. Overcooling or overheating can occur within a given zone if the system is not adjusted to respond to the load. Underventilation frequently occurs if the system is not arranged to introduce at least a minimum quantity (as opposed to percentage) of outdoor air as the VAV system throttles back from full airflow, or if the system supply air temperature is set too low for the loads present in the zone. In a VAV system, a VAV box in the occupied space regulates the amount of supply air delivered to the space, based on the thermal needs of the space. Malfunctioning VAV boxes can result in thermal discomfort and fail to prevent buildup of indoor air contaminants.

HYDRAULIC SYSTEMS

Modern hydraulics is defined as the use of confined liquid to transmit power, multiply force or produce motion. The transfer of energy takes place because a quantity of liquid is subject to pressure. A hydraulic system contains and confines a liquid in such a way that it uses the laws governing liquids to transmit power and do work. Pascal's Law states that pressure in a confined fluid is transmitted undiminished in every direction and acts with equal force on equal areas and at right angles to a container's walls. This principle serves as the foundation for all hydraulics systems operation. See Figure I-33.

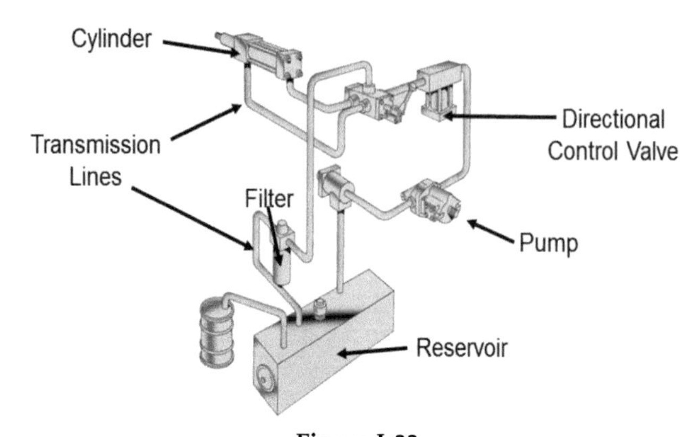

Figure I-33
Hydraulic System Components

Hydraulic systems contain the following key components:

Fluid—can be almost any liquid. The most common hydraulic fluids contain specially compounded petroleum oils that lubricate and protect the system from corrosion.

Reservoir—acts as a storehouse for the fluid and a heat dissipater.

Hydraulic pump—converts the mechanical energy into hydraulic energy by forcing hydraulic fluid, under pressure, from the reservoir into the system.

Fluid lines—transport the fluid to and from the pump through the hydraulic system. These lines can be rigid metal tubes or flexible hose assemblies. Fluid lines can transport fluid under pressure or vacuum (suction).

Hydraulic valves—control pressure, direction and flow rate of the hydraulic fluid.

Actuators (hydraulic cylinders)—convert hydraulic energy into technical energy to do work.

Warning! Warning! Warning!
High pressure fluid is present in operational hydraulic systems. Fluids under high pressure are dangerous and can cause serious injury or death. Do not make modifications, repairs or adjustments to any hydraulic system unless you are competent or working under competent supervision. If in doubt, consult a qualified technician or engineer.
Warning! Warning! Warning!

System Advantages

The advantages of hydraulic systems over other methods of power transmission are:

Simpler Design

In most cases, a few pre-engineered components will replace complicated mechanical linkages.

Flexibility

Pipes and hoses instead of mechanical elements virtually eliminate location problems.

Smoothness

Hydraulic systems are smooth and quiet in operation. Vibration is kept to a minimum.

Control

Control of a wide range of speed and forces is easily possible.

Cost

High efficiency with minimum friction loss keeps the cost of a power transmission at a minimum.

Overload Protection

Automatic valves guard the system against a breakdown from overloading. The main disadvantage of a hydraulic system is maintaining the precision parts when they are exposed to bad climates and dirty atmospheres. Protection against rust, corrosion, dirt, oil deterioration and other adverse environmental conditions is very important.

Circulatory System

Pipes and fittings, with their necessary seals, make up a circulatory system of liquid-powered equipment. Properly selecting and installing these components is very important. If improperly selected or installed, the result would be serious power loss or harmful liquid contamination. See Figure I-34.

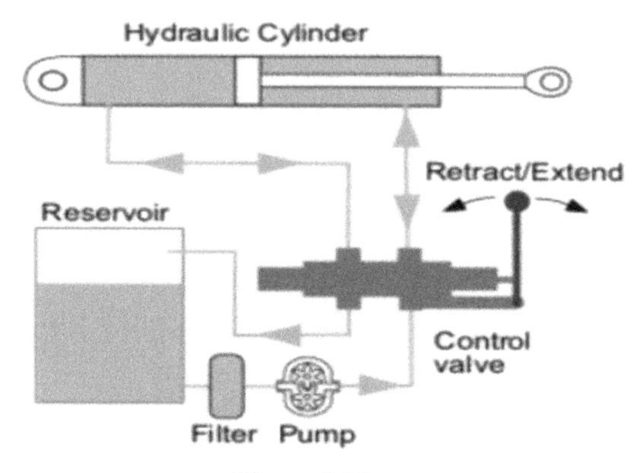

Figure I-34
Hydraulic Fluid Circulation

Following is a list of some of the basic requirements of a circulatory system:

- Lines must be strong enough to contain a liquid at a desired working pressure and any system surges.
- Lines must be strong enough to support the components that are mounted on them.
- Terminal fittings must be at all junctions where parts must be removed for repair or replacement.
- Line supports must be capable of damping the shock caused by pressure surges.
- Lines should have smooth interiors to reduce turbulent flow.

- Lines must be the correct size for the required liquid flow.
- Lines must be kept clean by regular flushing or purging.
- Sources of contaminants must be eliminated.

Filters, strainers, and magnetic plugs condition the fluid by removing harmful impurities that could clog passages and damage parts. Heat exchangers or coolers often are used to keep the oil temperature within safe limits and prevent deterioration of the oil. Accumulators, though technically sources of stored energy, act as fluid storehouses.

RECIPROCATING PUMPS

There are two basic types of pumps: positive displacement and centrifugal. The centrifugal types were covered earlier in the text. Reciprocating pumps are constant flow, positive displacement machines that make fluids move by trapping a fixed amount of it and forcing (displacing) that trapped volume into a discharge pipe. During operation, centrifugal pumps experience some internal leakage as the pressure increases, preventing a truly constant flow rate. On the other hand, positive displacement pumps can produce the same flow at a given speed (RPM), no matter what the discharge pressure. Reciprocating pumps consist of a "suction stroke" and a "delivery stroke." The suction stroke is where liquids are sucked into the pump, and the delivery stroke is where they are discharged from it. During the suction stroke, the piston is moved backward and the inlet valve opens, water enters and the exit valve will be closed. During delivery stroke the piston moves inward by forcing the water through the outlet valve, but the inlet valve remains closed. This cycle continues until the pump is stopped. See Figure I-35.

Positive displacement pumps are classified according to the mechanism used to move the fluid. Piston pumps, plunger pumps and diaphragm pumps are all classes of positive-displacement pumps, and are often used where a relatively small quantity of liquid is to be handled and where delivery pressure is quite large. Figure I-36 is an illustration of the (steam driven) bilge pumps we used on the ship during my Navy days. The direct-acting steam pump uses steam valves and valve gear as its method for reversing the direction of the pistons, in order to drive the pump.

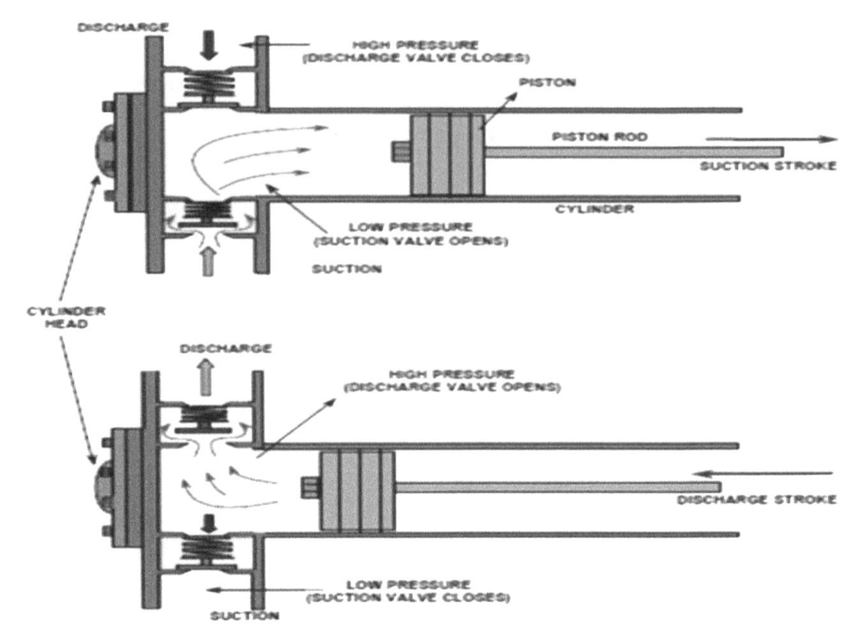

Figure I-35
Reciprocating Pump Action

Note: A positive displacement pump must never operate against a closed valve on the discharge side of the pump, because (as it continues to produce flow) the pressure in the discharge line increases until the line bursts, the pump is severely damaged, or both. Subsequently, placement of a relief or safety valve on the discharge side of the positive displacement pump is necessary.

Typically, a reciprocating pump consists of a cylinder with a reciprocating plunger in it. The suction and discharge valves are mounted in the head of the cylinder. In the suction stroke, the plunger retracts, and the suction valve opens, causing suction of fluid into the cylinder. In the forward stroke, the plunger pushes the liquid out of the discharge valve. With only one cylinder, the fluid flow varies between maximum flow (when the plunger moves through the middle positions) and zero flow (when the plunger is in the end positions). Positive displacement pumps are also categorized as single acting or double acting. A single acting pump is one which has one suction valve, one delivery valve and one suction and delivery pipe. It draws in the fluid in only one direction (suction stroke) and delivers it in a single stroke called the delivery

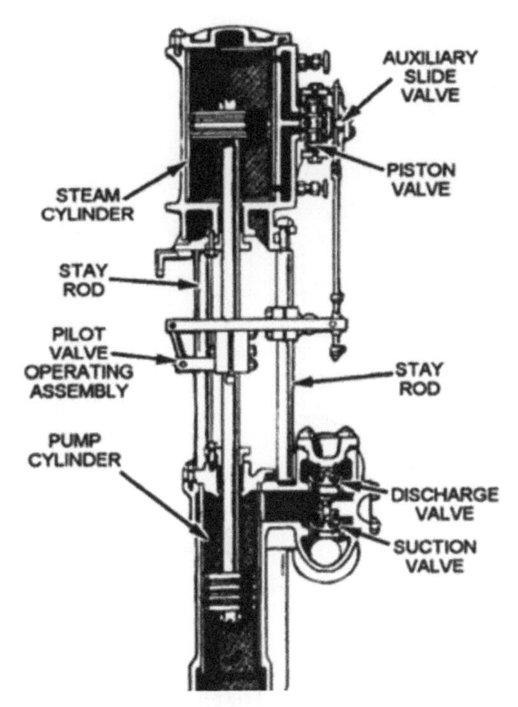

Figure I-36
Steam Driven Reciprocating Pump

stroke. A double acting pump is one which has two suction valves, two delivery valves and two suction and delivery pipes.

Diaphragm pumps are self-priming, reciprocating, positive displacement pumps that employ a flexible membrane instead of a piston or plunger to displace the pumped fluid. They operate via the same volumetric displacement principle described earlier. See Figure I-37.

Piping Considerations

Poor suction-side and discharge-side piping can result in excessive equipment vibration, noisy pump operation, load fluctuations, premature bearing failure, leaking flanges and wearing or failure of pump parts. Following proper piping practices can extend the life of the pumping equipment and its accessories.

Advantages and Disadvantages

Reciprocating pumps provide high suction lift and high outlet pressure, without the need for priming, and can be used for air service.

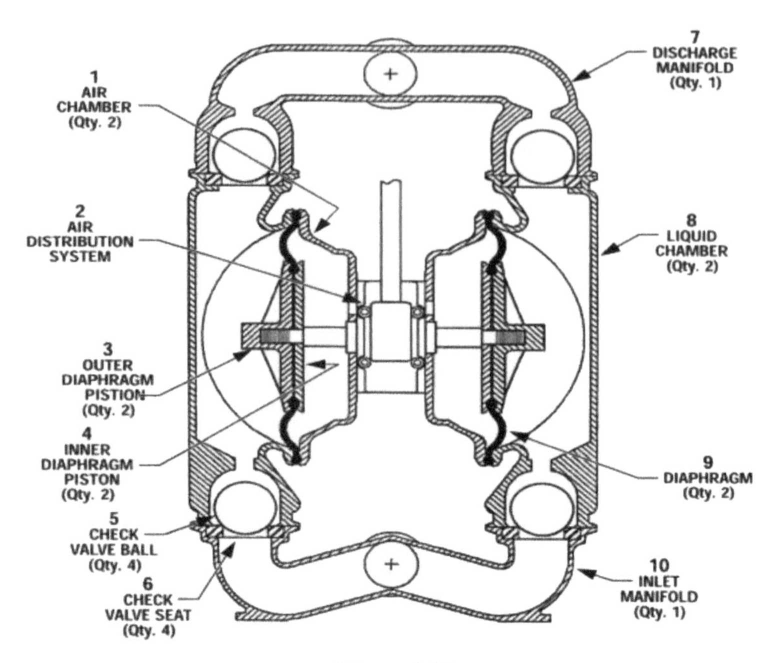

Figure I-37
Internal View of Diaphragm Pump

On the downside, they require a lot of maintenance (high wear and tear) and have a high initial cost.

STEAM DISTRIBUTION SYSTEMS

Steam is a versatile commodity used in physical plants for a wide variety of end uses. It is expensive to produce and behooves the plant to establish and maintain a comprehensive plan for ensuring the integrity of its application. There are four categories of concern regarding steam system components and ways to enhance their performance: its generation, distribution, recovery and end use. These areas represent the path that steam takes as it leaves the boiler and returns via the condensate return system. Figure I-38 depicts where the generation and recovery areas of the path occur.

There are many different end uses of steam. Mechanical drive and process heating make up the bulk of its use in the physical plant of

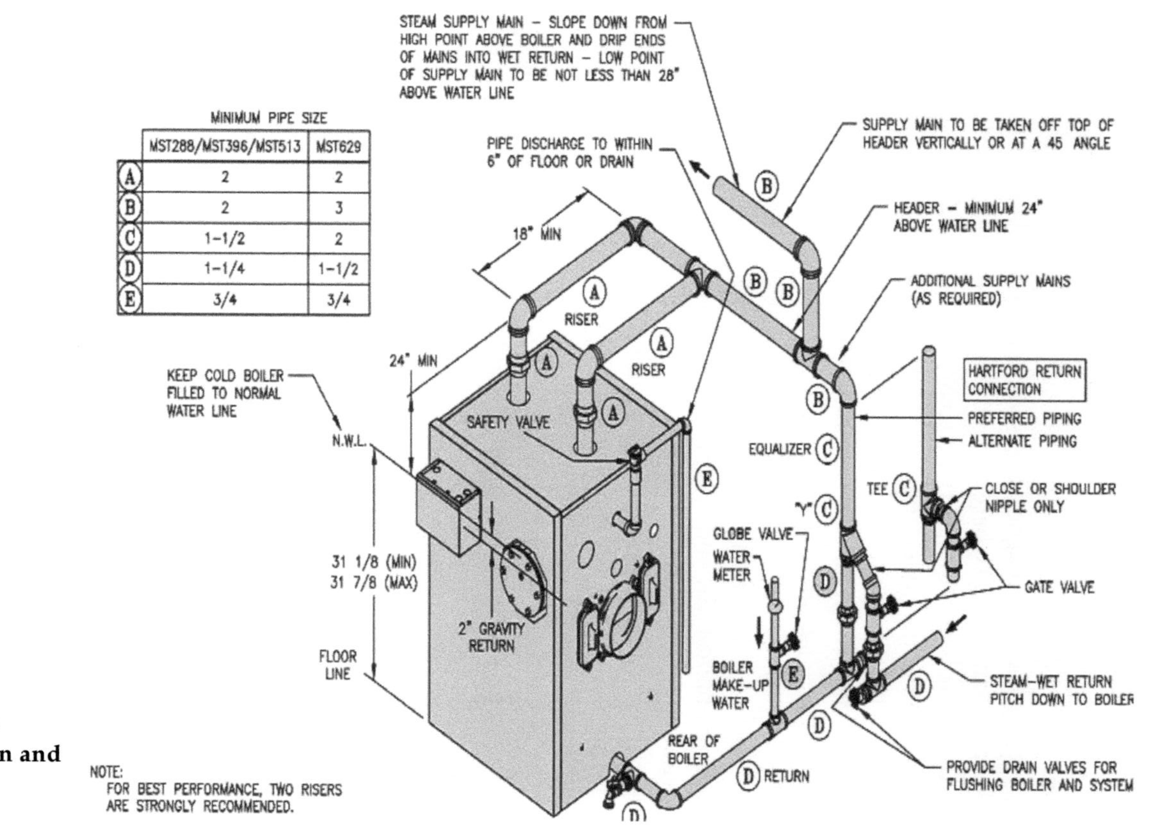

MINIMUM PIPE SIZE

	MST288/MST396/MST513	MST629
A	2	2
B	2	3
C	1-1/2	2
D	1-1/4	1-1/2
E	3/4	3/4

Figure I-38
Steam Generation and Recovery

buildings. Components of a typical steam distribution system commonly include: steam piping (where steam is transported between the boiler and end use devices, heat exchangers (where the steam transfers its latent heat to a process fluid), water softeners (where hardness minerals, such as calcium, magnesium, and iron are removed from the water supply), steam traps (where air and condensed steam are removed), condensate collection tanks (where condensate is collected, before being returned to the boiler), humidifiers (where steam is injected into an air source to increase its water vapor content), preheat and reheat air handling coils (where steam is used to heat an air supply), and steam valves (where steam flow is regulated to prevent over-pressurization and isolate equipment or system branches). See Figure I-39.

There are a wide range of steam system sizes, configurations, end-user applications and operating practices and several ways to improve system performance and identify opportunities for improvement. Improving steam system performance requires a systems approach to assessing the entire system, identifying opportunities and implementing the most feasible approaches. See Table I-3.

A systems approach analyzes both the supply and demand sides of the system and how they interact, shifting the focus from individual components to total system performance. Often, operators are so focused on the immediate demands of the equipment that they overlook the broader issue of how system parameters affect the equipment. A systems approach evaluates the entire system to determine how the requirements can be most effectively and efficiently served.

Note: This approach recognizes that system efficiency, reliability, and performance are related.

Inspect and Repair Steam Traps

In steam systems that have not been maintained for three to five years, 15% to 30% of the installed steam traps can fail, allowing live steam to escape into the condensate return system. In systems with a regularly scheduled maintenance program, leaking traps should account for less than 5% of the trap population. If your steam distribution system includes more than 500 traps, a steam trap survey will probably reveal significant steam losses.

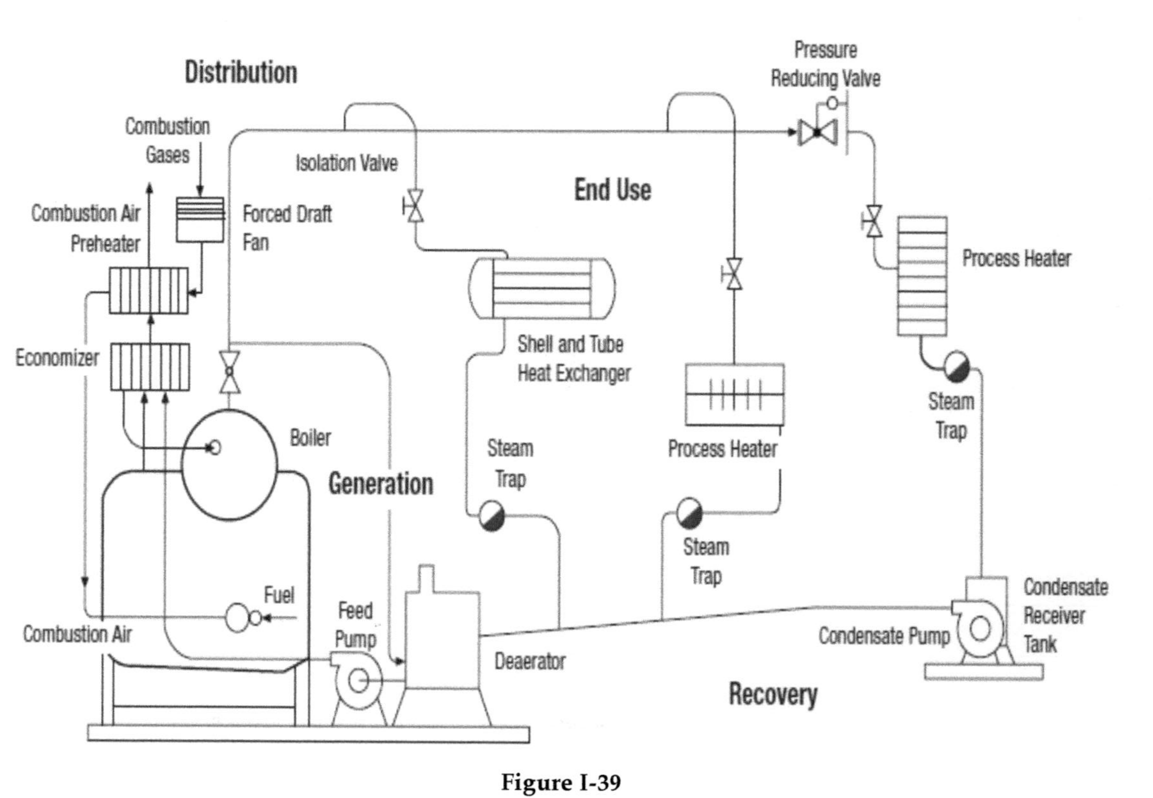

Figure I-39
Steam System Distribution Schematic

Table I-3
Operational improvements chart

Opportunity	*Description*
	Generation
Minimize excess air	Reduces the amount of heat lost up the stack, allowing more of the fuel energy to be transferred to the steam
Clean boiler heat transfer surfaces	Promotes effective heat transfer from the combustion gases to the steam
Install heat recovery equipment (feedwater economizers and/or combustion air preheaters)	Recovers available heat from exhaust gases and transfers it back into the system by preheating feedwater or combustion air
Improve water treatment to minimize boiler blowdown	Reduces the amount of total dissolved solids in the boiler water, which allows less blowdown and therefore less energy loss
Recover energy from boiler blowdown	Transfers the available energy in a blowdown stream back into the system, thereby reducing energy loss
Add/restore boiler refractory	Reduces heat loss from the boiler and restores boiler efficiency
Optimize dearator vent rate	Minimize avoidable loss of steam

	Distribution
Repair steam leaks	Minimizes avoidable loss of steam
Minimize vented steam	Minimizes avoidable loss of steam
Ensure that steam system piping valves, fittings, and vessels are well insulated	Reduces energy loss from piping and equipment surfaces
Implement an effective steam-trap maintenance program	Reduces passage of live steam into condensate system and promotes efficient operation of end-use heat transfer equipment
Isolate steam from unused lines	Minimizes avoidable loss of steam and reduces energy loss from piping and equipment surfaces
Utilize backpressure turbines instead of PRVs	Provides a more efficient method of reducing steam pressure for low-pressure services
	Recovery
Optimize condensate recovery	Recovers the thermal energy in the condensate and reduces the amount of makeup water added to the system, saving energy and chemicals treatment
Use high-pressure condensate to make low-pressure steam	Exploits the available energy in the returning condensate

Recommended Testing Intervals

- High pressure (150 psig and above)—weekly to monthly
- Medium pressure (30 to 150 psig)—monthly to quarterly
- Low pressure (below 30 psig)—annually

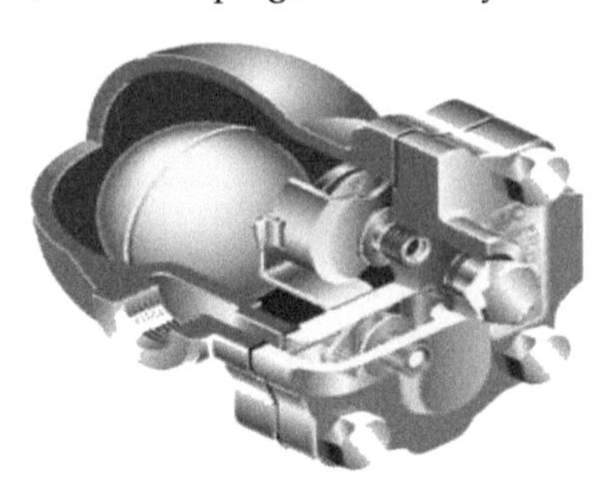

Figure I-40
Steam Trap

Insulate Steam Distribution and Condensate Return Lines

Steam distribution and condensate return lines that are not insulated are a constant source of wasted energy. Insulation can typically reduce energy losses by 90% and help to ensure proper steam pressure at plant equipment. Any surface over 120°F should be insulated, including boiler surfaces, steam and condensate return piping, and fittings. See Figure I-41.

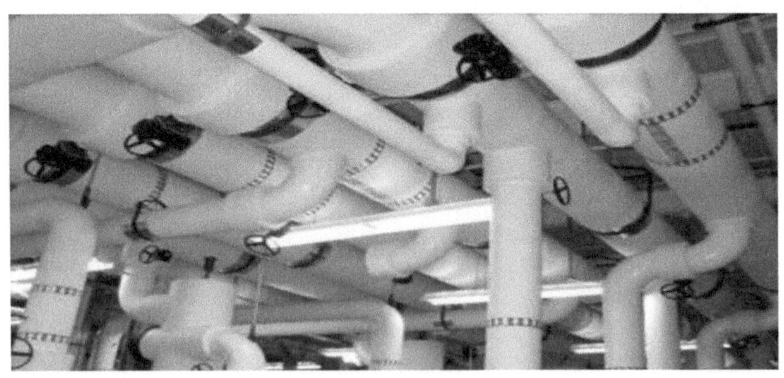

Figure I-41
Well Insulated Steam Distribution Lines

Insulation frequently becomes damaged or is removed and not replaced during steam system repairs. Wet or damaged insulation should be repaired or replaced immediately to avoid compromising its insulating value. Eliminate sources of moisture, prior to replacing insulation. Causes of wet insulation include leaking valves, external pipe leaks, or leaks from adjacent equipment. After steam lines are insulated, changes in heat flow can influence other parts of the steam system.

Return Condensate to the Boiler

When steam transfers its heat in a manufacturing process, heat exchanger, or heating coil, it reverts to a liquid phase called condensate. One way of improving your power plant's energy efficiency is to increase the condensate return to the boiler. Returning hot condensate to the boiler makes sense for several reasons. As more condensate is returned, less make-up water is required, saving fuel, make-up water, chemicals and treatment costs. Less condensate discharged into a sewer system reduces disposal costs. Return of high purity condensate also reduces energy losses due to boiler blowdown. Significant fuel savings occur as most returned condensate is relatively hot (130°F to 225°F), reducing the amount of cold make-up water (50°F to 60°F) that must be heated. Energy in the condensate can be more than 10% of the total steam energy content of a typical system.

Install Removable Insulation on Valves and Fittings

During maintenance, the insulation that covers pipes, valves and fittings is often damaged or removed and not replaced. Pipes, valves and fittings that are not insulated can be safety hazards and sources of heat loss. Removable and reusable insulating pads are available to cover almost any surface. The pads are made of a non-combustible inside cover, insulation material, and a noncombustible outside cover that resists tears and abrasion. Material used in the pads resists oil and water and has been designed for temperatures of up to 1600°F. Wire laced through grommets or straps with buckles hold the pads in place. See Figure I-42.

Figure I-42
Re-usable Steam Line Insulation Pads

Re-usable insulating pads are commonly used for insulating flanges, valves, expansion joints, heat exchangers, pumps, turbines, tanks and irregular surfaces. High-temperature piping and equipment should be insulated to reduce heat loss and improve safety. Insulating pads can be easily removed for periodic inspection or maintenance and replaced as needed.

Cover Heated, Open Vessels

Open vessels that contain heated liquids often have a high heat loss due to surface evaporation. Both energy and liquid losses are reduced by covering open vessels with insulated lids. It is assumed that the ambient air is dry with no wind currents. A fan pulling air over the uncovered tank could more than double the heat losses. See Figure I-43.

Heat Loss Detail

Eliminating internal heat gains will also result in electrical energy savings if the open tanks are located in conditioned space. Heat losses

Figure I-43
Well Insulated Condensate Receiver Tank

are a strong function of both wind velocity and ambient air humidity (wind velocities of 3 miles per hour will more than double the rate of heat loss from a tank). Radiation heat transfer is a secondary source of tank surface heat losses. Radiation losses increase from 90 Btu/hr-ft^2 at a liquid temperature of 110°F to 290 Btu/hr-ft^2 at 190°F.

TRANSFORMERS

Transformers convert electric power from one voltage level to another and are a crucial link in electric power distribution systems. They consist of two primary components: a core made of magnetically permeable material and a conductor, or winding, typically made of a low resistance material such as copper or aluminum. The conductors are wound around a magnetic core to transform current from one voltage to another. These essential electrical workhorses efficiently convert electricity to higher voltage for long-distance transmission and back down to low voltages suitable for customer use. See Figure I-44.

Utility transformers are high-voltage distribution transformers typically used by utilities to step down the voltage of electricity going into their customers' buildings. They are either mounted on an overhead pole or on a concrete pad. Distribution transformers are one of

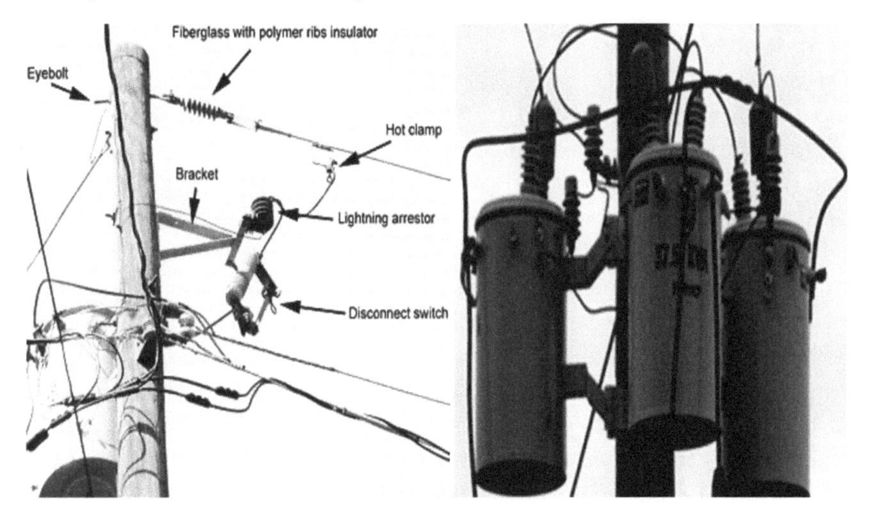

Figure I-44
Pole Mounted Transformers

the most widely used elements in the electric distribution system. They convert electricity from the high voltage levels in utility transmission systems to voltages that can safely be used in businesses and homes. Single-phase transformers supply single-phase service; two or three single-phase units can be used in a variety of configurations to supply three-phase service. Three-phase overhead transformer services are normally constructed from three single-phase units. Three-phase transformers for underground service (either pad-mounted, direct buried or in a vault or building or manhole) are normally single units, usually on a three- or five-legged core. Most commercial and industrial buildings require several low-voltage transformers to decrease the voltage of electricity received from the utility to the levels used to power lights, computers, and other electric-operated equipment. The distribution transformer normally serves as the final transition to the customer and often provides a local grounding reference. Most distribution circuits have hundreds of distribution transformers. Distribution feeders may also have other transformers—voltage regulators, feeder step banks to interface circuits of different voltages, and grounding banks.

> **Note**: A transformer's nameplate gives the kVA ratings, the voltage ratings, percent impedance, polarity, weight, connection diagram and cooling class. See Figure I-45 and Table I-4.

Transformer Cooling

Oil-filled Transformers

High voltage oil-filled transformers are bathed in transformer oil (a highly refined mineral oil that is stable at high temperatures). The transformer oil cools and provides part of the electrical insulation between internal live parts. By cooling the windings, the insulation will not break down as easily due to heat. To ensure that the insulating capability of the transformer oil does not deteriorate, the transformer casing is completely sealed against moisture ingress. Thus the oil serves as both a cooling medium to remove heat from the core and coil, and as part of the insulation system. Typically, transformers perform best at temperatures below 55°C above the ambient temperature. Large transformers to be used indoors must use a non-flammable transformer design liquid. In large power transformers, the oil-filled tank may have radiators through which the oil circulates by natural convection to improve cooling. Very large or high-power transformers (with capacities

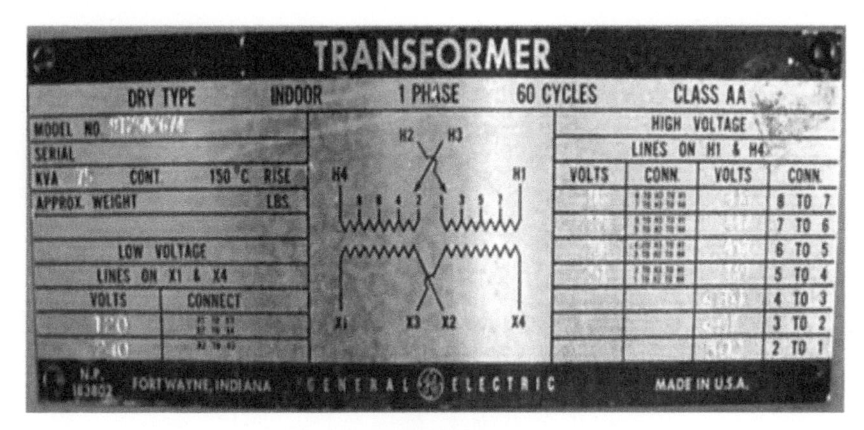

Figure I-45
Transformer Name Plate

Table I-4
Common Primary and Secondary Transformer Voltages

Single-phase Transformer		*Three-phase Transformer*	
Primary	*Secondary*	*Primary*	*Secondary*
240 x 480	120	208	208Y/120
120/208/240/277	120/240	240	240/120
480		480	480Y/277
		4,160	4,160
		13,800	

of millions of watts) may have cooling fans, oil pumps and even oil-to-water heat exchangers. See Figure I-46.

For large transformers used in power distribution or electrical substations, the core and coils of the transformer are immersed in oil which cools and insulates. Oil circulates through ducts in the coil and around the coil and core assembly, moved by convection. The oil is cooled by the outside of the tank in small ratings, and in larger ratings an air-cooled radiator is used. Where a higher rating is required, or where the transformer is used in a building or underground, oil pumps are used to circulate the oil and an oil-to-water heat exchanger may also be used. Heat is one of the most common destroyers of transformers. Operation at only 10°C above the transformer rating will cut transformer life by

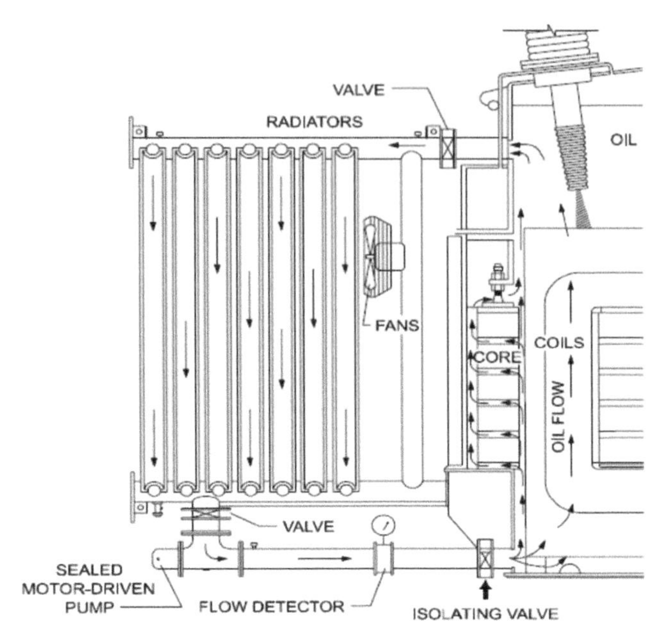

Figure I-46
Oil-filled Transformer

50%. Heat is caused by internal losses due to loading, high ambient temperature and solar radiation. It is important to understand how your particular transformers are cooled and how to detect problems in the cooling systems. ANSI and IEEE require the cooling class of each transformer to appear on its nameplate. In some transformers, more than one class of cooling and load rating are indicated.

Dry Type Transformers

Dry type transformers are used for both indoor and outdoor applications in hospitals, commercial buildings and anywhere that safe and dependable power are important considerations. They are not immersed in oil, but are cooled by air convection or by fans. It is important to keep transformer enclosures reasonably clean. It is also important to keep the area around them clear. Any items near or against the transformer impede heat transfer to cooling air around the enclosure. As dirt accumulates on cooling surfaces, it becomes more and more difficult for air around the transformer to remove heat. As a result, over time, the transformer temperature slowly rises unnoticed, reducing service life.

Transformer rooms and vaults should be ventilated. Portable fans (never water) may be used for additional cooling if necessary. A fan rated at about 100 cubic feet per minute (cfm) per kilowatt (kW) of transformer loss, located near the top of the room to remove hot air, will suffice. **Note**: These rooms/vaults should not be used as storage. Dry-type transformers, available at voltage ratings of 15 kV and below, are cooled primarily by internal air flow. The three principal classes of dry-type transformers are self-cooled (AA), forced-air cooled (AFA) and self-cooled/forced-air cooled (AA/FA). Self-cooled transformers require adequate room ventilation to ensure proper transformer cooling. Forced-air cooled transformers can be integrated into the facility energy conservation design by a heat recovery system. Ensure these transformers have minimum 0.3 meter (12-inch) spacing from combustible materials or have a fire-resistant, heat-insulating barrier. This requirement does not apply if the transformer is the non-ventilating type. See Figure I-47.

Construction Features

Transformer Taps

Taps are connection points along the transformer coil that effectively change the secondary voltage by changing the transformer turns ratio. Depending on the system conditions, the nominal secondary voltage might not satisfy the voltage requirements of the loads. They are provided on some transformers on the high-voltage winding to correct for high- or low-voltage conditions and still deliver full rated output voltages at the secondary terminals. General purpose transformers should be provided with several taps on the primary, to vary the secondary voltage. Standard tap arrangements are at 2-1/2% and 5% of the rated primary voltage for both high- and low-voltage conditions. See Figure I-48.

Noise

All transformers transmit sound due to vibration generated within the magnetic steel core. Depending on other nearby ambient noise, the transformer sounds might not be noticeable. In low ambient noise areas, the transformer sound can be noticed. Determine if noise rating is a required design consideration for the intended installation location. A transformer located in low ambient noise level areas should have a low decibel hum rating. In addition to the transformer noise rating, consider the following action to improve the generated sound level: Mount the

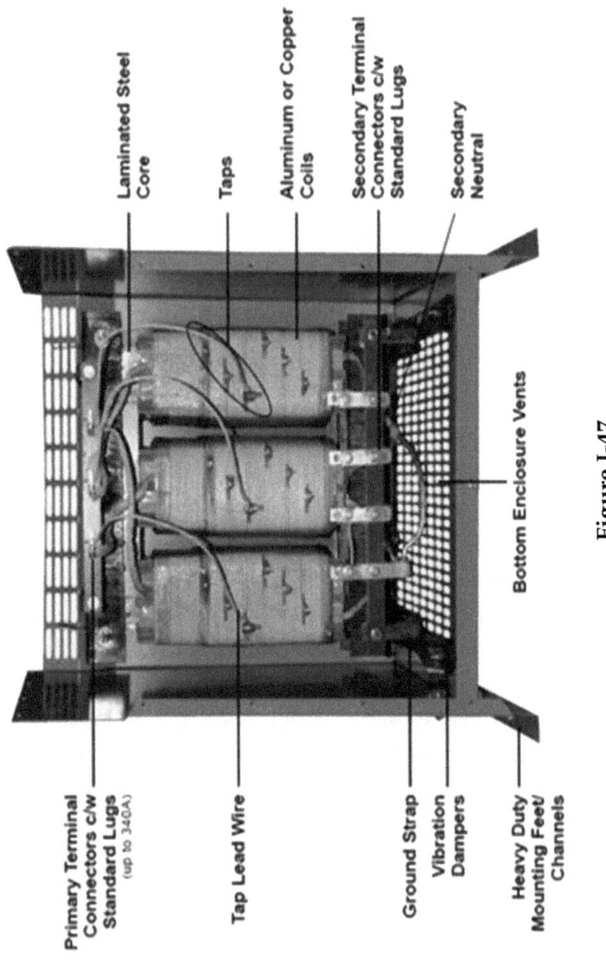

Figure I-47
Dry Type Transformer

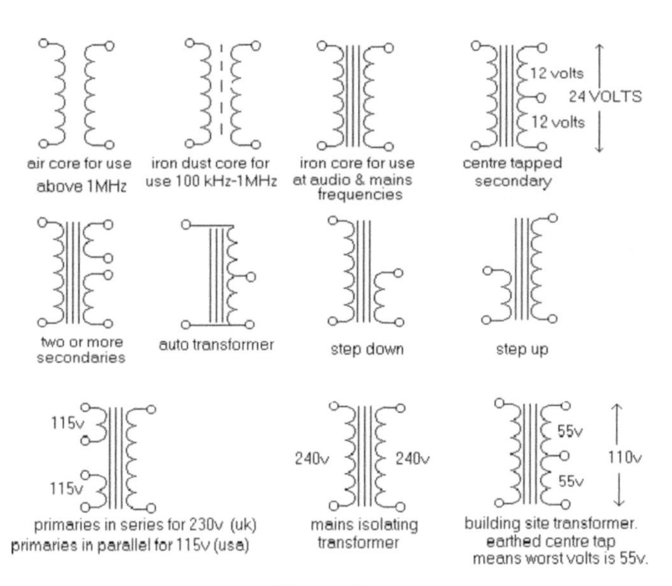

Figure I-48
Transformer Taps

transformer so that vibrations are not transmitted to the surrounding structure. Small transformers can usually be solidly mounted on concrete floors or walls. Flexible mounting will be necessary if the transformer is mounted to the structure in a normally low-ambient noise area. Use flexible couplings and conduit to minimize vibration transmission through the connection points. Locate the transformer in spaces where the sound level is not increased by sound reflection. For example, in terms of sound emission, the least desirable transformer location is in a corner near the ceiling because the walls and ceiling function as a megaphone. Transformers located within buildings where noise is of concern, such as hospitals or administrative facilities, must have a noise-level rating appropriate for the application. Vibration isolators should be provided to minimize sound transmission to the building's structural system.

> **Note**: The average sound level in decibels should not exceed the level specified in NEMA ST 20, Dry Type Transformers for General Applications, for the applicable kVA rating range.

Low Voltage Transformers

Transformers having a primary voltage of 600 volts or less for the supply of lower voltages should be of the self-cooled, ventilated dry type. Do not locate ventilated dry-type transformers in environments containing contaminants including dust, excessive moisture, chemicals, corrosive gases, oils or chemical vapors. Transformers should not be operated in parallel because the resulting interrupting duty requirements placed upon protective devices will increase the installation cost for such an arrangement. Also, transformers operated in parallel are subject to circulating currents, unless the impedances are carefully matched.

WATER DISTRIBUTION SYSTEMS

A water distribution system is the physical works that delivers water from the water source to the intended end point or user. Typically, this is achieved by way of pumps and motors, water mains, service pipes, storage tanks or reservoirs, and related equipment, in a closed system under pressure. See Figure I-49.

In buildings, that includes the water-service pipe, water meters, water-distributing pipes, and all necessary connecting pipes, fittings, pumps, control valves, and appurtenances. Proper maintenance of a water distribution system is important for providing high quality water

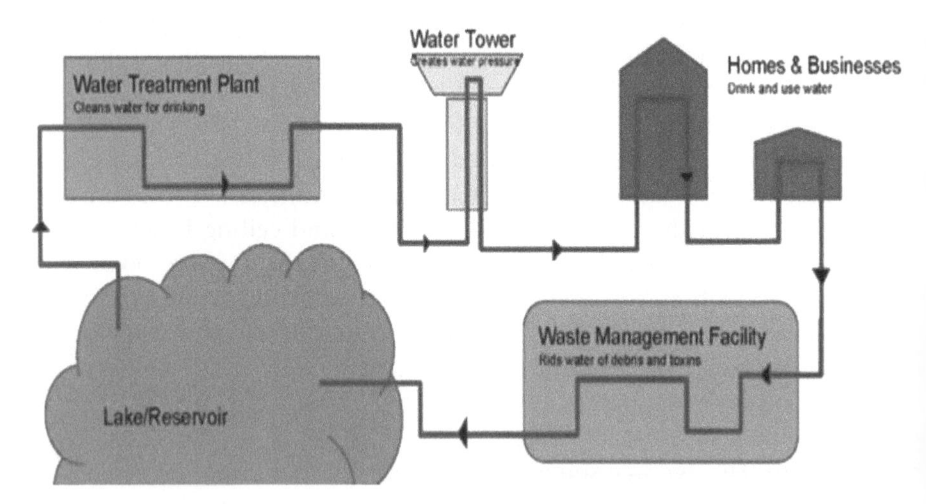

Figure I-49
Water Flow from Source to Service to Sewer to Source

to your customers, continuing operation in the event of an emergency, helping minimize property damage as a result of responding to an emergency, and helping prevent contamination events. A comprehensive preventive maintenance program can extend equipment life cycles and minimize problems related to minor or major equipment failures. The program can help you learn more about the features of the system's components, help you avoid problems with water quality and system breakdowns and reduce overall operating costs by preventing costly emergency repairs. Implementation of facility maintenance plans and schedules helps to assure that the system will be kept in optimum working condition. The preventative maintenance program should incorporate an asset assessment and management process that informs management about the quality of the physical facilities and the threats that weakened facilities might have on maintaining technical capability over time. Such a process includes identifying the physical assets of the water system, determining their age, condition and remaining useful life, developing a schedule of asset management or replacement, and calculating the costs of maintenance and/or replacement of assets in current and future years. See Figure I-50.

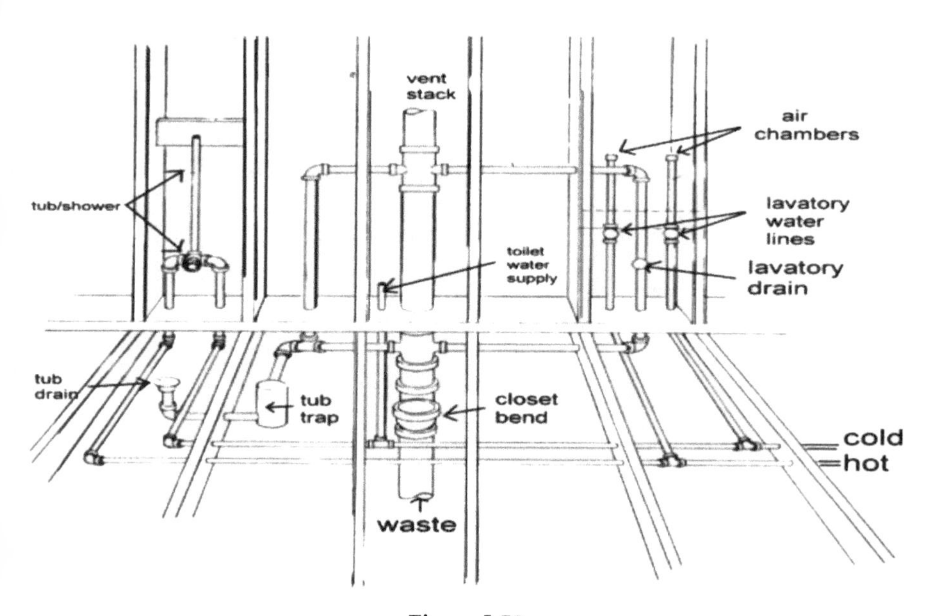

Figure I-50
Typical Appurtenances Found under the Floors and in the Walls

Metering & Backflow Prevention

There are five objectives for metering water supplies: to provide an incentive to conserve water (which protects water resources), to postpone costly system expansions, to save energy (and chemical costs), to enable utilities to better locate distribution losses, and to enable charges for water, based on use. Metering also helps to detect water leaks in the distribution network. See Figure I-51.

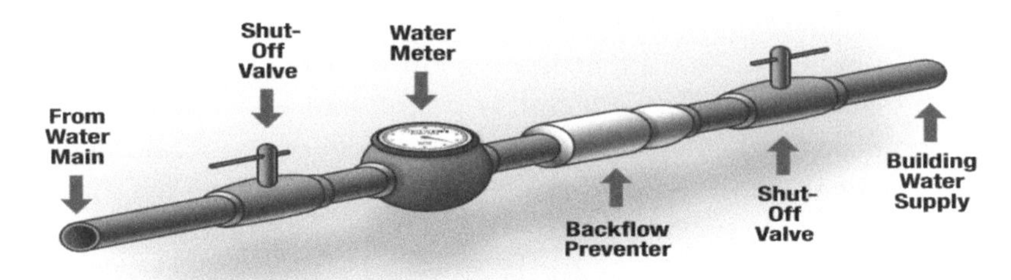

Figure I-51
Incoming Service Line

Backflow preventers are devices (normally found on incoming water system service lines) that are used to protect potable water supplies from contamination or pollution due to backflow. Water supply systems normally maintain a significant pressure, to enable efficient water flow to devices and processes downstream of the inlet line. Without a backflow preventer to protect the system, when water pressure in the line is disrupted (or shut off), the reduced pressure in the pipe may allow contaminated water from the soil, from storage or from other sources to be drawn up into the system. See Figure I-52.

Problems

Sudden changes in pressure can damage meters, such that many meters in water distribution systems may not be functional or accurate. Unless they are being replaced regularly, some types of meters become less accurate as they age and under-register consumption. Many types of meters also register air flows, which can lead to over-registration of consumption, especially in systems with intermittent supply, when water supply is re-established and the incoming water pushes air through the meters. Regular inspection and calibration is recommended.

Figure I-52

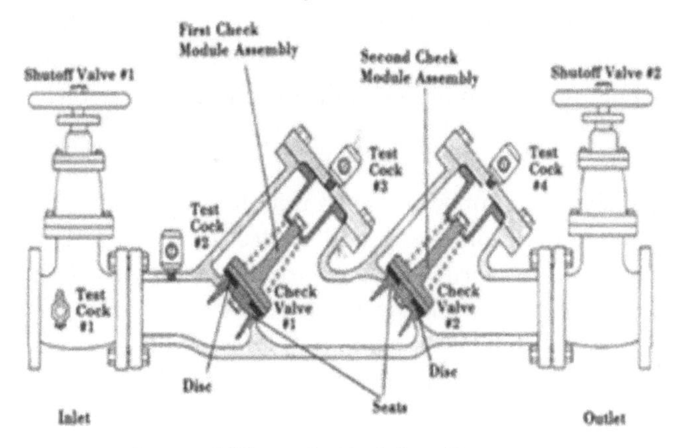

Internal View of a Backflow Preventer

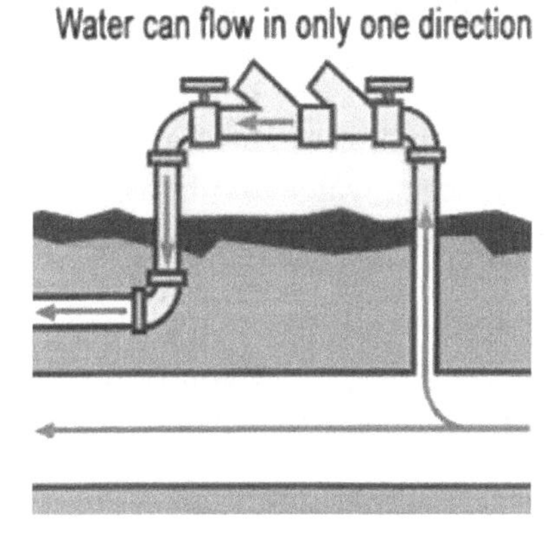

Backflow Preventer Action

Costs

The costs of metering include the investment costs to purchase and to install meters, as well as the recurrent costs to read meters and to issue bills based on consumption instead of bills based on monthly flat fees.

Potable Water

Potable water is water which is fit for consumption by humans and other animals. In reference to its intended use, it is also called drinking water. Water may be naturally potable, as is the case with pristine springs, or it may need to be treated in order to be safe. Not all water supply systems are used to deliver drinking water. Systems used for purposes such as heating and cooling (of buildings) and firefighting (sprinkler systems) operate in much the same way as systems for drinking water, but the water need not meet such high standards of purity.

Legionella

One of the main concerns when maintaining water quality in potable systems is the monitoring and control of the *Legionella* bacillus. As bacteria found in warm water environments, *Legionella* is virtually impossible to completely eradicate from hot water tanks and cooling towers. The goal, instead, is to contain it. Thus, facility and maintenance managers need to be diligent in their planning, implementation and documentation of procedures to keep *Legionella* under control. Areas most susceptible to *Legionella* proliferation are pipe valves and joints, bottoms of holding tanks and faucet aerators. Your goal should be to avoid water temperatures within the ideal *Legionella* growth range. Thus, pipes as well as holding tank water should be outside of the range of 20°-50°C (68°-122°F). But this alone is not enough for an effective *Legionella* management plan. Regular temperature checks, water sampling, and tests of material collected at faucets, water fountains, shower heads and other fixtures, can help keep a containment plan working.

Piping Problems

The piping system design can contribute to *Legionella* bacteria growth. *Legionella* bacteria incubate or grow in warm stagnant water where there is bio-film or media for it to grow on. Such conditions exist in existing piping systems, but can be controlled by removing dead-legs (capped off branches), avoiding washers/gaskets made of natural rubber, replacing heavily scaled faucets/shower heads, avoiding the use of faucet aerators, avoiding water hammer air cushion chambers and minimizing cool zones in hot water tanks.

Leaks/Conservation

Aging water systems increase water losses. Leaks can be major sources of water loss in a distribution system. The costs of losing large volumes of water through leaks or blow-off valves may be alleviated by investing in repair or reconstruction of water lines to current design standards. The goal of conserving water to save money requires a vision beyond the up-front costs associated with the initial investment in water-efficient fixtures and equipment. The Federal Energy Policy Act of 1992 addressed water conservation by mandating water efficiency standards for indoor water fixtures. The EPA has also developed a program called Water Sense that enables consumers to easily identify water-efficient products that do not sacrifice performance or quality. The program website provides directories of service providers and partnered manufacturers, retailers and distributors of water-efficient products. The Water Sense program is similar to the Energy Star program for energy-efficient products.

Section II

Associated Accessories and Mechanisms

AIR FILTER MAINTENANCE

Air filters are the first line of defense in protecting heat transfer surfaces and ensuring building air quality. Reductions in air flow cause air handling units (AHUs) to work harder than they should, making the units less energy efficient than they could be. Air flow efficiency constantly changes when filter media get loaded down with particles such as dust (mites), pollen, mold spores, hair, bacteria, lint and other substances, until they become clogged and flow of air is greatly reduced. Having the right filters in place, and performing proper periodical maintenance on air filtering systems, improves IAQ and maximizes the investment. Selection of filters and filter media is based on factors such as:

- Air volume and cleanliness required in the space
- Types, sizes and concentrations of contaminants
- Required air-filter efficiency
- Spatial accessibility for installation
- Replacement and storage costs
- Life-cycle costs (including disposal)

Characteristics

The three main characteristics of proper filter selection and performance are efficiency, resistance to airflow, and dust-holding capacity. Efficiency measures the ability of the filter to remove particles from the air stream. Resistance refers to the static-pressure drop across the filter at a given face velocity. Dust-holding capacity defines the amount of dust an air filter can hold when it operates at a specified airflow rate to some maximum resistance value. Evaluating filter types requires data on efficiency, resistance during filter loading and dust-holding capacity at various pressure drops. These items can directly affect the fan's ability to move air at varying resistances, which can lead to higher fan brake horsepower. Uniform air velocity across all filters in a rack extends filter life and assures airflow and filter integrity during system operation and filter loading. This situation extends the filter's performance life and cuts energy use.

Classifications

Filters and filter media are classified by efficiency, method and use. In accordance with the ASHRAE standard (52-76), efficiency classes include low (20 percent or less of dust spot efficiency), medium (more than 20 percent, less than 90 percent), and high (more than 90 percent, or more than or equal to 95 percent DOP test—0.3 micron smoke). Methods classifications consist of inertial impingement (where particles are trapped on media fibers by being impinged by the force created by their weight and high velocity; and interception (where particles too small and too light in weight to be impinged are removed most economically by interception, which occurs when their path is altered by air molecules after their velocity is slowed by passing through media.

Filter media used in interception is pleated and usually of finer fiber than inertial impingement media; electronic agglomeration or electrostatic precipitation (electronically charging dust particles that will collect on oppositely charged plates). Usage is described as pre-filter (both coils and final filtration systems are protected by removing larger dust particles and contaminants upstream, usually very near outside air dampers. Pre-filters are usually:

- panel-type using inertial impingement method
- general ventilation and air conditioning (includes low, medium, and high efficiency ranges, dry, viscous-treated, washable, and disposable media types, in a variety of frames, construction, and arrangement
- downstream high efficiency particulate air filter beds: (designed to meet requirements of Department of Health, Education, and Welfare Resources Publication No. 76-4000 for Hospital and Medical Facilities), and
- industrial (includes special media and filter arrangements to remove lint, press ink mist, and non-atmospheric contaminants).

Protecting the most expensive filters with pre-filters makes practical sense, and extends filter replacement time, while potentially reducing housekeeping and maintenance activities. Installing a manometer or draft gauge across each filter bed marked for clean and dirty resistance is a good method that allows technicians to visually check the filter loading.

Figure II-1
Pleated Pre-filter

Construction

Protecting more expensive filters with a pre-filter maximizes the investment and extends filter replacement time, while potentially reducing housekeeping and maintenance activities. This can be accomplished with the use of panel media with coarse fibers (usually called throwaway filters). The medium most common in disposable panel filters is glass fiber, although polyester fiber is also used. The fiber is treated with an oil or adhesive spray to create a viscous impingement medium. This is usually found in an efficiency range of 20 percent or less (ASHRAE average with atmospheric dust). Other types of filters and filter media include:

HEPA Filter

HEPA filters have a high ratio of filter area to face area and fine fibers with tightly controlled spacing, closely pleated. Performance characteristics typically include an efficiency of 99.97% on 0.3 micron size particles at an initial resistance of 1.0"/w.g. (water gauge) and final resistance of 2.0" to 3.0," depending upon construction. See Figure II-2.

Figure II-2
HEPA (box) Filter

Roll Media

Typically furnished as a pre-filter or upstream filter for a higher efficiency filter system, the roll media filter may be used where 25% average efficiency will suffice. The usual roll media installation is an automatically advanced roll of adhesive-coated glass fiber filter media that is fed into the air stream and rerolled after it has collected its dust load. Roll media filters come in a variety of media constructions and resulting efficiency characteristics that allow their use as pre-filters. They are used alone as the principal filtration device or to retain agglomerated dust from electrostatic precipitators. See Figure II-3.

Electronic Filters

Two types are generally found in HVAC systems: (1) the agglomerator-type with disposable collection media, or (2) the precipitator/collector-type, which requires washing and renewal of collector plates, automatically or manually. The principle of electrostatic precipitation is the same—dust and smoke particles are given positive or negative static charges by the electrostatic field set up by the charged ionizing wires

Figure II-3
Roll Media

and the grounded struts. Charged particles then enter the collecting section, which is made up of alternately positive charged and grounded plates. The charged particles are attracted to and held by the oppositely charged plates.

Inspection

Four important items require attention in filter installations; i.e., support of the filter frames, air leakage around the frames, the fit of the filter media in the frames and the condition of the media itself. If sufficient attention isn't given to these items, the following results may be experienced:

- Unsupported filter frames (lacking rigidity) can collapse as the filters load.
- Cracks around frames and duct walls can allow unfiltered air to leak.
- Incorrectly installed or out-of-plumb frames will also allow air to bypass.
- Bag filters that are pinched shut or not fully open and extended will reduce efficiency. See Figure II-4.

Figure II-4
Bag Filter

Summary

The benefits of a good air filtration service program are improved air quality, energy savings (through reduced resistance to air flow, permitting lower fan speed and resulting in lower horsepower), enhanced building appearance, and reduced labor time cleaning coils and ductwork.

BELTS, DRIVES AND PULLEYS

Aside from direct coupling, belt drives are the most cost-effective and reliable means of power transmission (chain drives require constant lubrication, and gear drives are prone to mechanical problems). However, their reliability is contingent on the level of care they receive on a regular basis. Belts and drives should be subjected to an effective maintenance program which includes these elements:

- effective inventories
- regular belt drive inspections
- proper belt installation procedures
- belt product knowledge
- belt drive performance evaluations
- troubleshooting guidelines
- proper storage techniques
- maintaining a safe working environment

Preventive maintenance should be performed regularly. Critical drives may require a quick visual and hearing inspection every week or two. On most drives, that cursory procedure can be performed once a month. Every 3 to 6 months, the driven units should be shut down for complete servicing (thorough inspection of belts or pulleys and other drive components). Figure II-5 shows a representative sampling volume of the deficiencies commonly found during the inspection and preventive maintenance process.

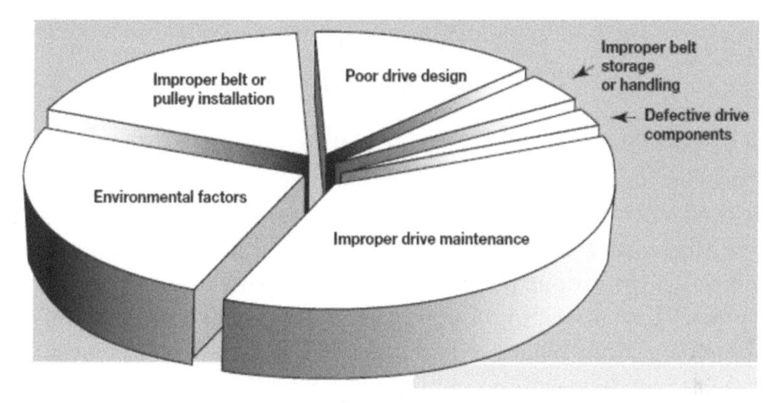

Figure II-5
Common Drive, Belt and Pulley Deficiencies

Belt Drives

A V-belt drive is a very efficient and forgiving method of power transmission. As such, proper alignment and belt tension are extremely important and can make a huge difference in mean time between repairs (MTBR). Belt drive maintenance involves performing a close inspection of the belts and of each sheave and its grooves to look (and feel) for cracks, chips, or excessive groove wear, and check for proper contact between the belts and the sheaves. Deficiencies found should be corrected before proceeding with the remainder of the preventive maintenance process. Good sheave alignment will increase efficiency by reducing premature wear or failure of belts, pulleys and bearings. See Figure II-6.

Misalignment

Misalignment consists of three types (that can co-exist in combination); i.e. vertical angularity, horizontal angularity and axial offset. Prior to beginning a sheave alignment, it is a good idea to try to deter-

Figure II-6
Belt Drive Pulley

mine (if possible) the cause of your belt or sheave failure, and correct it to prevent unnecessary re-occurrence. The cause of failure could be associated with poor drive maintenance (improper belt tension, poor sheave alignment), environmental factors (sunlight, harsh temperature fluctuations), improper installation (wrong belts/sheaves, belts pried on by force), or operating factors (overload, shock-load). **Note**: Only combine belts from the same manufacturer, and preferably use a factory-matched set. Once alignment has been established, inspect the belt(s) for any noticeable defect (cracking, gouges or crumbling) and signs of slippage (glazing). Notice if and where the belt is worn. This may be a good indication of what type of misalignment or other problem might be in play. Place the new belts into the sheave grooves, reposition the sheaves to rough alignment, and check that the belts are properly seated within their grooves.

Belt Tensioning

Note: The friction between the belt and the wheel is affected by centrifugal force which tends to lift the belt off the wheel. This increases the likelihood of slippage. The friction between the belt and wheel may be increased by the shape of the belt. A vee section or round section belt in a vee groove will grip better than a flat belt and is less likely to slip. See Figure II-7.

Pulley drive systems depend on friction to enable the belt to grip the wheel and pull it around with it. To enable this, the belt must be tensioned, even when the wheels are stationary. After the sheaves have been aligned and the belts have been replaced, they must be properly tensioned. Incorrect tension (as well as misalignment) will adversely

Figure II-7
Multi-pulley V-belt Assembly

affect the life of the belts and the efficiency of the drive as a whole. The force (tension) in a pulley belt increases with torque and power. The maximum power that a pulley system can transmit is ultimately limited by the strength of the belt material. Tension the belts until the force required for this deflection equals the belt manufacturer's maximum recommended force values for the specific belts you are using. Using a matched set of belts and having a good alignment is essential in achieving this goal. It is tricky to move the driver to slacken or tighten the belts without changing the alignment! See Figure II-8.

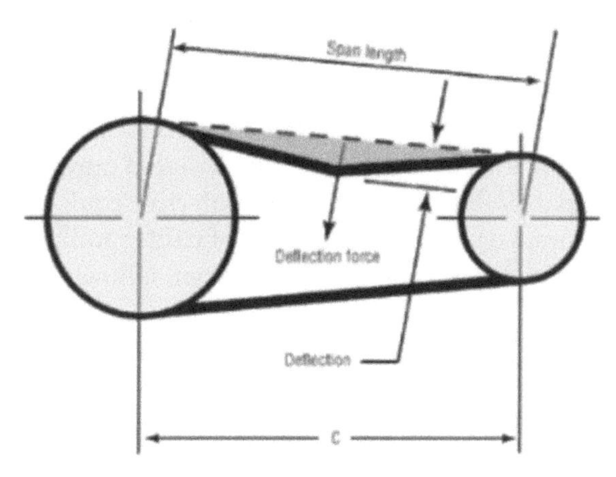

Figure II-8
Belt Deflection Requirement between Pulleys

Note: Change a belt anytime undue wear is detected. Belts should be stored in a cool, dry place with no exposure to direct sunlight or heater drafts. Do not hang belts from a single peg; this may damage the tensile members and distort the belt, over time. Preferably, hang them on two pegs, or better yet, pile them on shelves. Coil long belts, and don't make the piles too big or heavy (to avoid distorting the bottommost belts.

After belt installation, run the machines for 2 hours to allow the belts to stretch and seat themselves properly in the grooves. The belts must then be re-tensioned to the recommended values. Run the machines at least 72 hours but not more than 10 days, and re-tension once again, this time to the manufacturer's recommended force values for used belts. During start-up, look and listen for unusual noise or vibration. It is a good idea to shut down the machine and check the bearings and motor. It is extremely important that pulleys be installed and aligned properly. If they feel hot, the belt tension may be too tight, or the bearing may be misaligned or improperly lubricated. See Figure II-9.

Operator Safety

Operator safety is of paramount importance when servicing belt drives. Take the following precautions to ensure the safety of the operator or mechanic; always follow manufacturer's recommendations:

- Utilize proper lock-out/tag-out procedures.
- Have only trained personnel work on the belt drives.
- Always turn the equipment off before beginning work.
- If possible, remove fuses from the motor control center (MCC).
- Place components in neutral position to avoid movements.
- Wear proper clothing and personal protective equipment.
- Keep the areas around the drive free of clutter and parts.
- Clean oil and debris from floors to prevent falling.

Machine Guards

Always keep drives properly guarded. A well designed and installed guard:

- Completely encloses the drive.
- Has grills or vents for good ventilation.

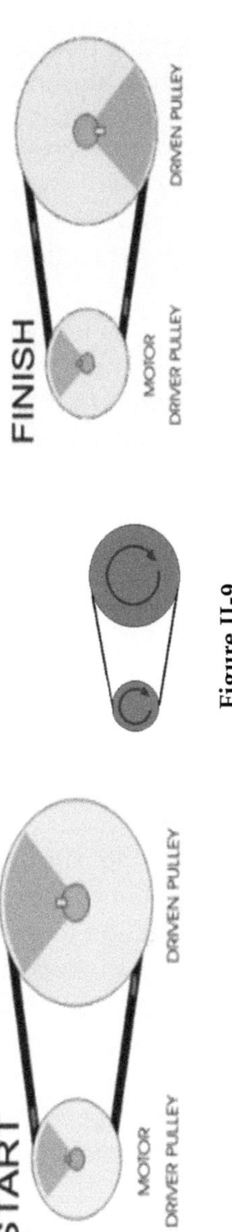

Figure II-9
Pulley Ratio Showing Rotation

- Is equipped with an automatic shutoff device.
- Has accessible inspection doors or panels.
- Can easily be removed and replaced if damaged.
- Protects the drive from weather, debris and damage.
- Keeps operating personnel out of harm's way.

Check guards for wear or possible damage. Look for signs of wear or rubbing against drive components. Clean them to prevent their becoming insulated and closed to ventilation. Clean off any grease or oil that may have been spilled onto the guard from over-lubricated bearings.

Preventive Maintenance

Experience with your own equipment will be the best guide to how often you need to inspect the belt drives. High speeds, heavy loads, frequent start/stop conditions, extreme temperatures and drives operating on critical equipment will mean more frequent inspections. The following factors will influence the frequency of drive inspection:

- temperature extremes
- drive operating speed
- drive operating cycle
- critical nature of equipment
- environmental factors
- accessibility of equipment

Preventive maintenance actions include: turning off power to the drive, locking out the control box, placing all machine components in a safe (neutral) position to avoid accidental movement, removing and inspecting the guard to check for wear (rubbing), cleaning the guard, inspecting the belts and pulleys for wear or damage, inspecting bearings/shafts/motor mounts, belt tension, pulley alignment, power sources and unit operation. Adequate periodic care and repair of these devices will ensure years of dependable service. Some minor modifications that might improve their operation include:

- increasing pulley diameters
- increasing the number of belts
- utilizing wider or premium belts
- adding vibration damping

- improving guard ventilation
- reducing operating temperature
- replacing worn or bent pulleys
- keeping pulleys properly aligned
- re-tensioning newly installed belts

COMPRESSED GAS CYLINDERS

A gas cylinder is a pressure vessel used to store gases at above atmospheric pressure. High-pressure gas cylinders are also called bottles. Although they are sometimes colloquially called "tanks," this is technically incorrect, as a tank is a vessel used to store liquids at ambient pressure and often has an open top. The use, storage and handling of compressed gas cylinders presents two types of hazards. First is the chemical hazard associated with the cylinder contents (corrosive, toxic, flammable, etc.). Second is the physical hazard represented by the cylinder being under pressure. Mishandled cylinders may rupture violently, release their hazardous contents, or become dangerous projectiles. See Figures II-10 and II-11.

Note: Gas cylinders are often color coded, but the codes are not standard across different jurisdictions and sometimes are not regulated. Cylinder color cannot safely be used for positive product identification; cylinders have labels to identify the gas they contain, and the label alone should be used for positive identification.

Figure II-10
Compressed Gas Cylinders

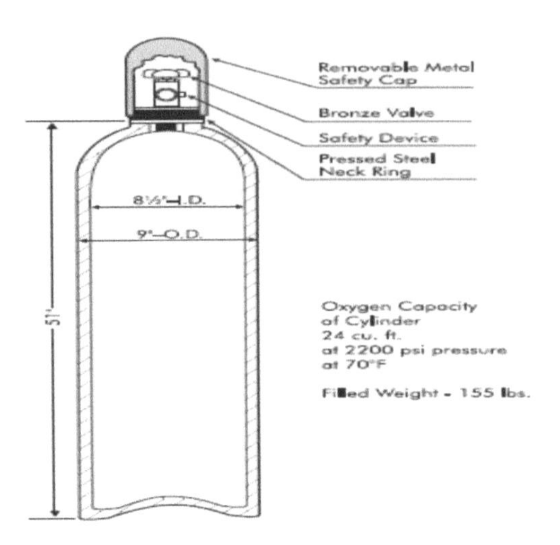

Figure II-11
Cylinder Construction

Regulatory Guidelines

Department of Transportation Classifications

Your compressed gas cylinders will have one or more of the hazardous materials placards shown in Figure II-12. The United States Department of Transportation (US DOT) in Title 49 Section 173 of the United States Code of Federal Regulations (49 CFR 173) requires the use of hazardous materials placards when shipping compressed gases. These hazardous materials placards are intended to indicate the general hazards associated with the contents of the gas in the cylinder. For complete hazardous material information, refer to the Material Data Safety Sheet (MSDS).

Hazard Identification System

In Section 704 of the National Fire Code, the National Fire Protection Agency (NFPA) specifies a system for identifying the hazards associated with materials. The hazard identification signal is a color-coded array of four numbers or letters arranged in a diamond shape. An example is shown in Figure II-13. You will see hazard diamonds like this on trucks, storage tanks, bottles of chemicals, and in various other places around campus. The blue, red, and yellow fields (health, flammability, and reactivity) all use a numbering scale ranging from 0 to 4. A value of zero means that the material poses essentially no hazard; a rating of

Figure II-12
Hazardous Materials Placards

four indicates extreme danger. The fourth value (associated with white) tends to be more variable, both in meaning and in what letters or numbers are written there. The NFPA diamond is designed to give general hazard information for chemicals. Figure II-13 spells out those values.

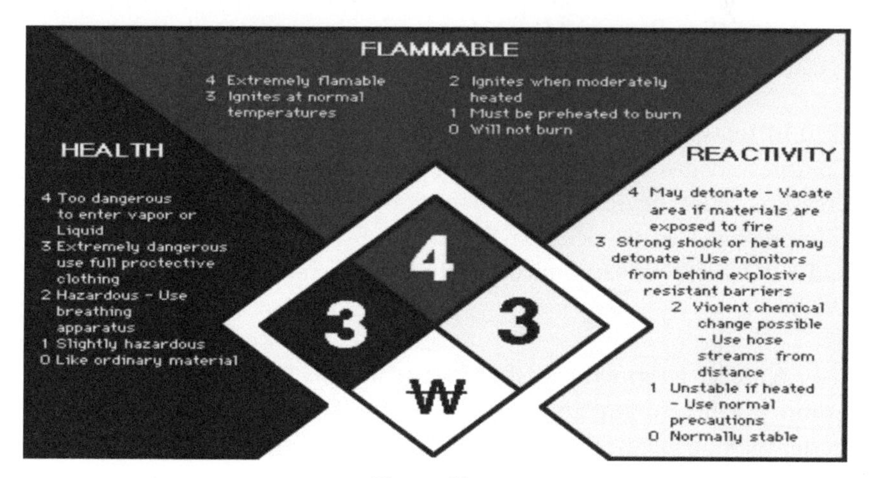

Figure II-13
National Fire Protection Agency (NFPA) Diamond

The NFPA 704 diamond is divided into four colored quadrants that provide information about the material stored inside the given facility. These quadrants create a standard system for identifying hazards. Each color in the diamond represents a specific type of hazard.

- Blue represents a health hazard
- Red represents flammability
- Yellow represents reactivity
- White provides information about special precautions

OSHA Requirements

1910.101(a)—"Inspection of compressed gas cylinders." Each employer shall determine that compressed gas cylinders under his control are in a safe condition to the extent that this can be determined by visual inspection. Visual and other inspections shall be conducted as prescribed in the Hazardous Materials Regulations of the Department of Transportation (49 CFR parts 171-179 and 14 CFR part 103). Where those regulations are not applicable, visual and other inspections shall be conducted in accordance with Compressed Gas Association Pamphlets C-6-1968 and C-8-1962, which is incorporated by reference as specified in Sec. 1910.6.

1910.101(b)—"Compressed gases." The in-plant handling, storage, and utilization of all compressed gases in cylinders, portable tanks, rail tank cars, or motor vehicle cargo tanks shall be in accordance with Compressed Gas Association Pamphlet P-1-1965, which is incorporated by reference as specified in Sec. 1910.6.

1910.101(c)—"Safety relief devices for compressed gas containers." Compressed gas cylinders, portable tanks, and cargo tanks shall have pressure relief devices installed and maintained in accordance with Compressed Gas Association Pamphlets S-1.1-1963 and 1965 addenda and S-1.2-1963, which is incorporated by reference as specified in Sec. 1910.6

Note: It is imperative that the facility managers of each building group (facility) maintain a complete inventory of ALL gas cylinders under their auspices. The information tracked should include:

- Types, numbers and sizes of cylinders
- Manufacturer and vendor ownership
- Contents and pressures
- Quantity and volume
- Proximity of content substances
- Manifolds, valves and regulators

- Areas where gas is utilized
- Specific hazard classes
- Cylinder age and delivery dates
- Date of last hydrostatic test
- Handling, transportation and storage procedures
- Emergency procedures

CRITICAL SPARES

Critical spares are those items identified as being required to be on-hand due to potential catastrophic equipment failure or as a consequence of unacceptable interruptions of the facility's operations or overall mission, the impact being so great that it would be more costly in downtime and lost production than to purchase and store. By providing critical spare parts supply programs, costly downtime as a result of equipment failure is minimized and productivity and workflow increased. These items provide risk mitigation for managing and limiting the potential for catastrophic failures. Not all spares need to be on site. Not all spares need to be on order. But those "critical" spares that can shut us down when they are not available need to be covered by a plan that minimizes downtime and expense for our operations and ensures customer satisfaction.

Identifying Stock

Excessive or inadequate stocking of maintenance parts and materials can result in financial over-expenditures, operational interruptions and equipment downtime. For the management of maintenance spares and materials to be cost-effective, a thorough understanding of the facility's systems and equipment nuances must be acquired and addressed. Differing models are available to arrive at optimum stocking levels considering myriad scenarios. The focus here is on a simpler approach. Our first step to arrive at optimum inventory levels is to classify the stock as either:

- Critical spares
- Non critical spares
- Consumables

Critical spares are those without which machines can't be operated, have high failure rates, long procurement lead times, and no substitutes, or are used in mission-critical devices. Non-critical spares are those that are highly reliability, can be easily or readily made or purchased, can be substituted for, are available "off the shelf" as standard parts or that will allow the device to operate (albeit, less efficiently). Consumables such as gaskets, belts, chemicals or small components are items commonly utilized for performing preventive maintenance on equipment.

Determinants

A critical spares program should provide a proactive, repeatable approach to the identification, management, stocking, and visibility of managed items that might negatively impact operational readiness. Typically, three primary factors are taken into account in spare parts criticality analyses: functional location, mean time between failures (MTBF) and lead time. The total criticalities of different spare parts at the same functional location are differentiated on the basis of their mean time between failures and delivery times. But, none of the three can guide a stocking decision by itself. However, combining them with other factors and operations attributes, into a total spare-part criticality score can provide a comprehensive and rational result. The criticality of the spare to the overall operation is then used as the major criterion to set safe minimum levels for stock-holding policy. Economic considerations are used to recommend the maximum level. Focus should be on the consideration and weighting of chosen "contributing" factors such as age, cost, condition, duty, failure patterns, commonality of parts, predictability, impact of downtime, where used, substitutability, warrantees in effect or other chosen (preferred) impact factors. Point values should be assigned to each of the applicable factors for use in a subsequent "criticality" analysis where they will be applied to operational characteristics or attributes chosen for the criticality analysis These include:

- Customer requirements
- Mission impact
- Planned utilization
- Delivery lead times
- Safety and environment
- Reliability (MTBF)

- Single points of failure
- Repair histories
- Asset replacement value
- Regulations

Note: To identify the significance of a "spare part" to the operations, each attribute is then weighed against the contributing factors.

Gauging Risk Levels

As the result of a spare parts criticality analysis, every recommended spare part gets an individual criticality score, which facilitates comparisons between spare parts. The benefit of this type of comparative spare parts analysis is that it provides a rational basis for the selection of the items stocked, and also makes it easier to identify items that a supplier can stock, that can be carried on-site (in general stores) or kept close by the equipment to which they are allotted. Assigning a level of risk to each enables determination as to what actions are taken regarding stock items.

The risk priority associated with each type of failure determines the appropriate corrective actions. Risk priorities might include:

- *Low Risk and Low Consequence*—The risk is accepted not to carry the item in stock.

- *Medium Risk and High Consequence*—The risk of failure is sufficiently high to invest in on-site/warehoused spares.

- *High Risk*—Spares should be readily available and located near the devices to which they are allotted.

Once we understand the meaning behind the criticality analysis model, it becomes a tool used to develop the asset management program. To make asset management a plant-wide process, all critical characteristics that are common throughout the campus, across all assets, critical or not, must be considered. For instance, if "Mission Impact" is commonly critical, then the facility manager may need to consider equipment redundancy plans. If "Spares Lead Time" is found to be particularly critical, a materials management improvement program might be initiated.

Program Principles

Spare parts inventory management shares many traits with standard inventory management. Efficient spare parts inventory management plays a critical role in reducing costs and maximizing customer service. The institutional knowledge required for implementing and executing an effective spare parts inventory management program includes understanding projected consumption, calculating system failure costs, knowing cost impact of out-of-stocks, determining cost-reduction and in-stock improvement, and calculating costs of expedited orders.

ELECTRICAL SWITCHGEAR INSTRUMENTATION

Electrical switchboards are large free-standing, single-panel, frame or assembly of panels that are replete with switches, buses, instruments, overcurrent and other protective devices mounted on the face or back or both and are accessible from both the front and rear of the enclosure. As a general rule, switchboards are located at the "Main Electrical Disconnect" where electrical service first enters a building, and they are considered the primary distribution points within a building. Switchboards function the same way as the typical home panel box, subdividing and distributing power to a number of individual branch circuits. If the switchboard acts as a service entrance (as in the accompanying graphic), it may also contain a main disconnect and overcurrent protection device. In this case, the switchboard feeds lighting panel boards and HVAC circuits. This scenario is typical of those found in commercial building applications. See Figure II-14.

In larger installations, each branch circuit (panel board) will have its own individual disconnect and overcurrent protection device. Panel boards, can be single panels or a group of panel units designed for assembly in the form of a single panel. They include buses, automatic overcurrent devices, and are equipped with or without switches for the control of light, heat, or power circuits. They are designed to be placed in a cabinet or cutout box placed in or against a wall or partition and accessible only from the front. They are commonly used as a means of electrical distribution between main or sub-distribution points and individual connected loads. Different types of panel boards are commonly available, as follows:

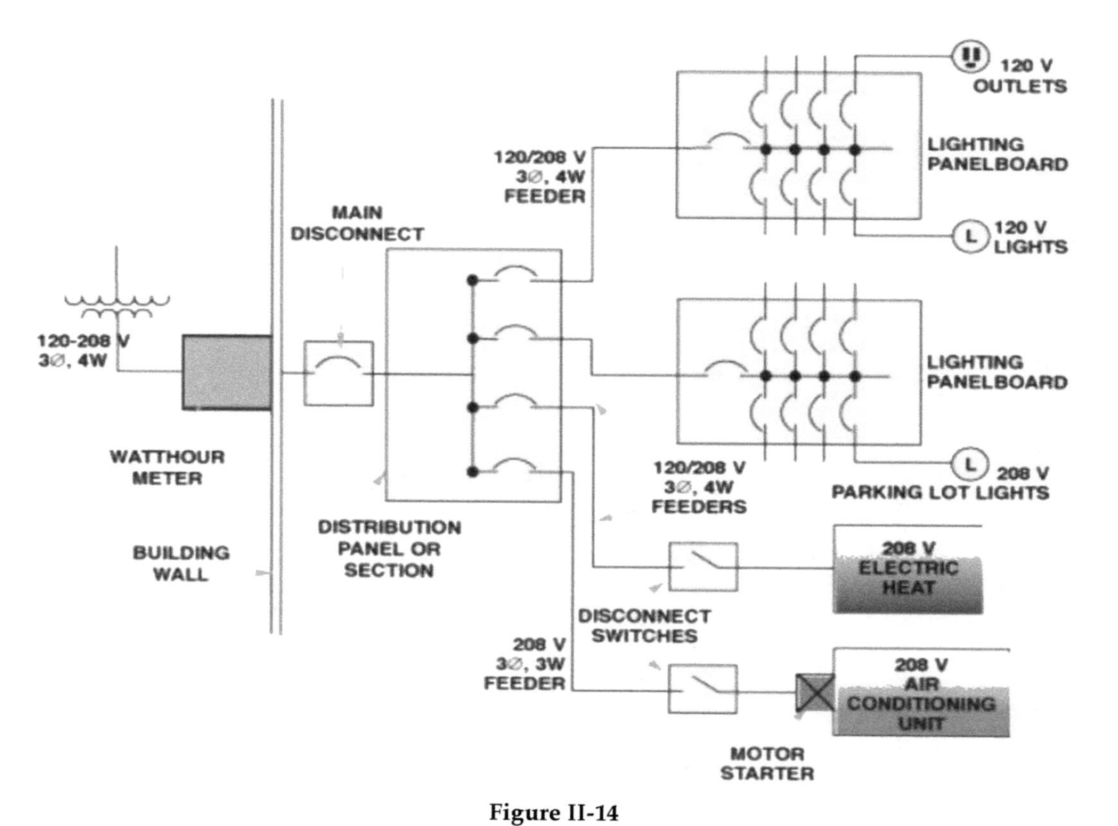

Figure II-14
Building Service Entrance Showing Branch Circuits

Service equipment panel boards—available for loads up to 1,600 amperes. These panel boards usually contain up to six breakers (or fused switches or fusible bolted pressure switches) connected to the incoming mains.

Feeder distribution panel boards—usually contain circuit overcurrent devices rated at more than 50 amperes to protect sub-feeders to smaller branch circuit panel boards.

Load center panel boards—normally rated up to 1,200 amperes at 600 volts or less. Similar to feeder distribution panelboards, these contain three-phase control and overcurrent devices for motor or power circuit loads. See Figure II-15.

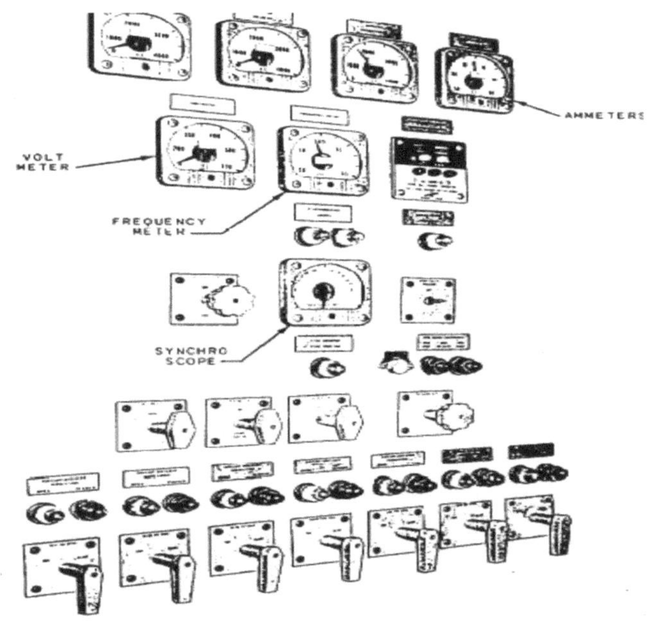

Figure II-15
Load Center Instrument Panel

General Criteria

In an electric power system, switchgear is the combination of electrical disconnects, switches, fuses or circuit breakers used to control, protect and isolate electrical equipment. Switchgear is used both to de-energize equipment to allow work to be done and to clear faults downstream. This type of equipment is important because it is directly linked to the reliability of the electricity supply.

> **Notes**: Do not route piping containing liquids, corrosive gases, or hazardous gases in the vicinity of switchgear unless suitable barriers are installed to protect the switchgear from damage in the event of a pipe failure. Do not locate switchgear where foreign flammable or corrosive gases routinely and normally are discharged. Do not use switchgear enclosure surfaces as physical support for any item unless specifically designed for that purpose. Do not use enclosure interiors as storage areas unless specifically designed for that purpose.

Plant Instrumentation

An instrument is a device that measures and/or regulates physical quantity/process variables such as electrical values, flow, temperature, level or pressure. Instrumentation is "the design, construction, and provision of instruments for measurement, control, etc.; the state of being equipped with or controlled by such instruments collectively." The application of instrumentation is the practice of utilizing devices to measure and control variables within a process. Instrumentation can also refer to hand-held devices that measure some desired variable. Instrument technicians and mechanics specialize in troubleshooting, repairing and maintaining individual instruments and instrumentation systems. Their trade is closely intertwined with those of electricians, pipefitters, power engineers and engineering companies. Tradesmen employ myriad devices in their daily work. Output instrumentation includes devices such as solenoids, valves, regulators, circuit breakers and relays. These devices control a desired output variable and provide either remote or automated control capabilities.

In the electrical field, instrumentation is used to measure many parameters (physical values), including electrical power, voltage, amperage, continuity, frequency, inductance, impedance, capacitance and resistivity, among others. Measurement instruments have three traditional classes of use:

- monitoring of operations
- control of operations
- engineering analysis

While these uses appear distinct, in practice they are less so. All measurements have the potential for decisions and control. As mechanic, you observe or monitor operating equipment and take the necessary steps to detect malfunctions and prevent damage to the equipment. The word *monitor* means to observe, record, or detect an operation or condition using instruments. By reading and interpreting the instruments, you can determine whether the equipment or the system is operating within the prescribed range.

Electrical Indicating Instruments

Electrical indicating instruments (meters) are used to display information that is measured by some type of electrical sensor. Although meters display units such as pressure or temperature, the meters on the control console are, in fact, DC voltmeters. The signal being sensed is conditioned by a signal conditioner. Electrical values, such as power and current, are measured and displayed at service switchboards.

Voltmeters

DC and AC voltmeters determine voltage the same way. Both measure the current that the voltage is able to force through a high resistance. This resistance is connected in series with the indicating mechanism or element. Voltmeters installed in switchboards and control consoles all have a fixed resistance value. Portable voltmeters, used as test equipment, usually have a variable resistance. For both installed and portable voltmeters, resistances are calibrated to the different ranges that the meters will display. The normal range for the switchboard and electric plant meters is 0 to 600 volts. See Figure II-16.

Figure II-16
Voltmeter

Ammeters

Ammeters are used to measure the amount of

current passing through a conductor. Different types of ammeters are used to measure either AC or DC. Ammeters that are designed specifically to indicate AC will also measure DC, but with a lower degree of accuracy. Ammeters must be connected in series with the circuit to be measured. For this reason, installed ammeters are constructed so that they do not handle the current that passes through the conductor being measured. Since ammeters cannot handle the high switchboard current, the switchboard ammeters operate through current transformers. This arrangement isolates the instruments from the line potential. In its secondary, the current transformer produces a definite fraction of the primary current. This arrangement makes it possible for you to measure large amounts of current with a small ammeter. See Figure II-17.

Figure II-17
Ammeter

Frequency Meters

Frequency meters measure cycles per rate of AC. Operating frequency in the United States is 60 Hz (Hertz). In Europe and other parts of the world it is 50 Hz. A frequency meter may have a transducer that converts the input frequency to an equivalent DC output. The transducer is a static device that has two separately tuned series-resonant circuits, which feed a full-wave bridge rectifier. A change in frequency causes a change in the balance of the bridge. This causes a change in the DC output voltage. See Figure II-18.

Figure II-18
Frequency Meter

Note: Often, emergency power (stand-by) generators are tied into the main switchboard through either manual or automatic transfer switches. Both kilowatt meters (used for balancing the electrical load) and synchroscopes (for paralleling the load for transfer) are often employed when operating and manipulating these devices manually.

Kilowatt Meters

Power is measured by computing values of current, voltage, and the power factor. Kilowatt meters automatically take these values into account when they are measuring kilowatts (kW) produced by a generator. Kilowatt meters are connected to both current and potential transformers so they can measure line current and voltage. The amount of power produced by a generator is measured in kilowatts. Therefore, when balancing the electrical load on two or more generators, you should make sure the kW is matched. See Figure II-19.

Loss of the kW load is the first indication of a failing generator. For example, if two generators are in parallel, and one of the two units is failing, you should compare the kW readings. Normally, the generator with the lowest kW would be the failing unit. There is one case, however, where this is not true. During an over-speed condition, both units increase in frequency, but the failing unit is the one with the higher load.

Figure II-19
Kilowatt Meter

Synchroscopes

Synchroscopes are power factor meter connections that measure the phase relationship between the generator and bus bar voltages. The moving element is free to rotate continuously. When the two frequencies are exactly the same, the moving element holds a fixed position. This shows the constant phase relationship between the generator and bus-bar voltages. When the frequency is slightly different, the phase relationship is always changing. When this happens, the moving element of the synchroscope rotates constantly. The speed of rotation is equal to the difference in frequency; the direction shows whether the generator is fast or slow. The generator is placed on line when the pointer slowly approaches a mark. This mark shows that the generator and bus-bar voltages are in phase. See Figure II-20.

Figure II-20
Synchroscope

Before connecting a three-phase generator to bus-bars already connected to one or more generators or electrical feeds, you must make sure that certain conditions prevail. The phase sequence for both generator and bus bars must be the same. A synchroscope is used to find out if the following required conditions have been met for the generator and busbar:

- voltages must be the same
- frequency must be the same
- voltages must be in phase

Note: They must reach their maximum voltages at the same time and their frequencies must remain constant for an appreciable amount of time; therefore, when connected, they will oppose excessive circulation of current between the two electrical sources.

Power Factor Meter

The power factor of an AC electrical power system is defined as the ratio of the real power flowing to the load to the apparent power in the circuit (the ratio of power dissipated over input: the ratio of the actual power dissipated in an electrical system to the input power of volts multiplied by amps). A power factor meter monitors that value at between 0 and 1. At a power factor of unity, one potential coil current leads and one lags the current in Phase B by 30°; thus, the coils are balanced in the shown position. A change in power factor will cause the current of one potential coil to become more in phase and the other potential coil to be more out of phase with the current in another phase, so that the moving element and pointer take a new position of balance to show the new power factor. See Figure II-21.

Figure II-21
Power Factor Meter

ELECTRICAL GROUNDING AND LIGHTNING PROTECTION

Electrical protection ("grounding" or "earthing") is the most important element of modern electrical systems and lightning protection designs. The purpose of electrical grounding is to provide protection for electrical equipment and personnel by eliminating the possibility of dangerous or excessive voltages and currents. Without proper electrical protection, personnel are at higher risk of shock; equipment operation can be negatively impacted by ambient electrical noise; and electronic and electrical equipment is at risk of damage from voltage and current surges. Power system ground faults always find a path to ground. Therefore, functional electrical protection systems must be in place that will safely channel and dissipate this errant electrical energy to prevent personnel. Grounding in electrical systems is the connection of bonded metallic circuit elements through engineered, low impedance paths to Earth. The ground is the negative or minus part of a circuit where the voltage is zero—such as the minus side of batteries and power supplies. The point may or may not be connected to Earth. A wire that is directly connected to Earth is also at ground. Proper grounding of circuits and devices helps prevent component breakage and circuit damage and fires, as well as helping to prevent injury or death to the person using them. See Figure II-22.

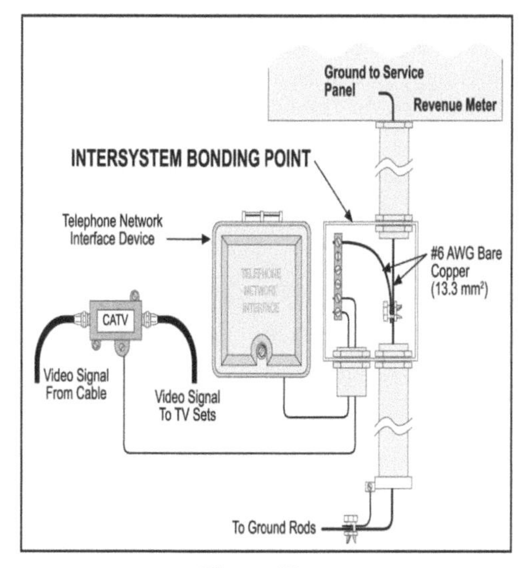

Figure II-22
Electrical System Grounding

Electrical grounding is important because it provides a reference voltage level (called zero potential or ground potential) against which all other voltages in a system are established and measured. An effective electrical ground connection also minimizes the susceptibility of equipment to interference, reduces the risk of equipment damage due to lightning, and eliminates electrostatic buildup that can damage system components by draining away any unwanted buildup of electrical charge. When a point is connected to a good ground, that point tends to stay at a constant voltage, regardless of what happens elsewhere in the circuit or system. The Earth, which forms the ultimate ground, has the ability to absorb or dissipate an unlimited amount of electrical charge. A well designed grounding system can: provide a safe and effective path for dissipation of fault currents, lightning strokes, static charges and EMI and RFI signals; mitigate the harmful effects of lightning strikes and electrical system fault conditions; lessen the chance of personnel injury due to fault currents or lightning strikes; and reduce the likelihood of electrical system damage due to lightning. See Figure II-23 (opposite).

Proper grounding is essential to ensure normal operations. All electrical currents seek the path of least resistance as a basic principle of circuit operation. Grounding provides this path of least resistance where, once a current is generated by a power supply, it will flow through all circuit components in an effort to reach the ground point, similar to water flowing downhill. Done improperly, this natural current flow results in electrical arching and electrical signals cross coupling or interfering with other circuits. In cases where there are more than one equipment grounding conductors leading into a junction box, these have to be spliced or joined, inside the box. If metal junction boxes are used, the grounding conductors from each device must be connected to the box with a grounding screw. See Figure II-24.

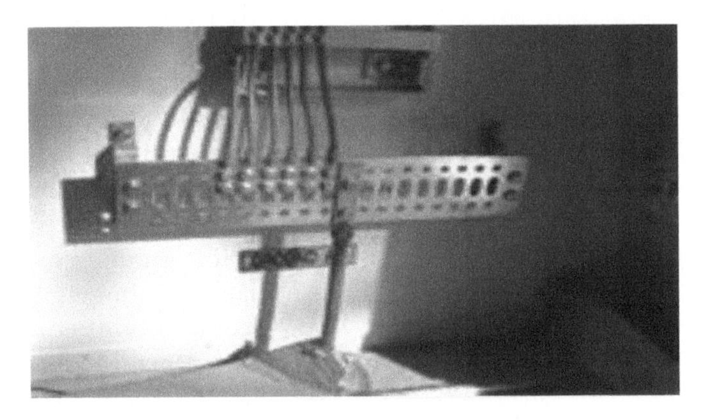

Figure II-24
System Grounding Manifold

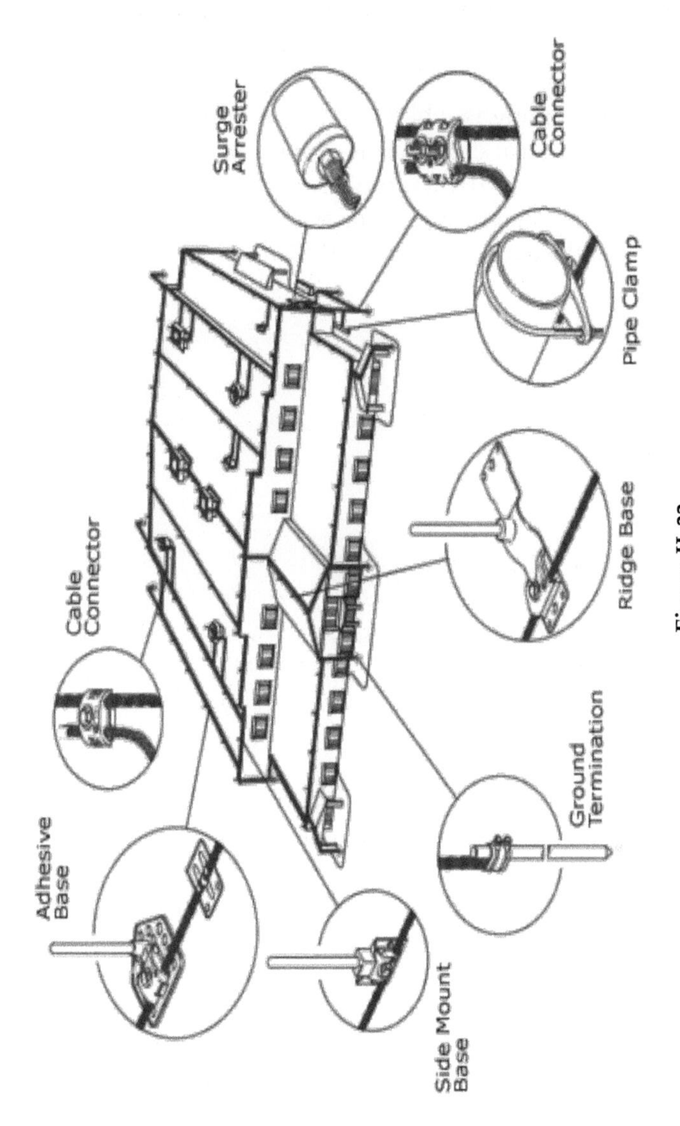

Figure II-23
Electrical System Grounding Utensils

Note: It is advisable to provide a bare wire conductor as a ground connection; particularly in electronic circuits. Naked wire helps to dissipate any accumulated charge spread to surfaces it touches, thus preventing a buildup of a charge on the conductors.

Earth Ground

Electrical grounding is where a point in an electrical circuit is at zero voltage. In electrical engineering, *ground* or *earth* can refer to the reference point in an electrical circuit from which other voltages are measured, or a common return path for electric current, or a direct physical connection to the earth. See Figure II-25.

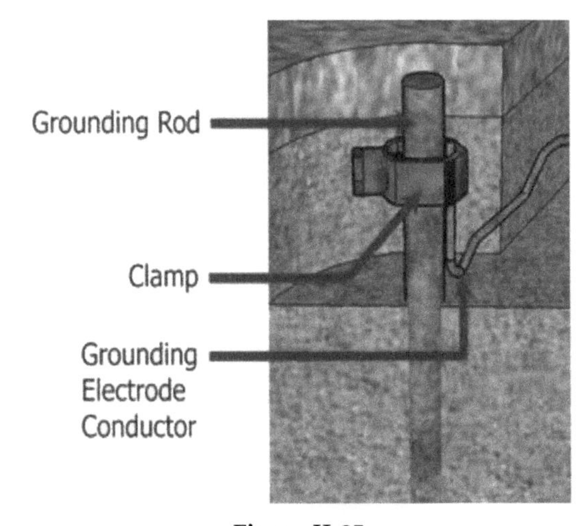

Figure II-25
Earth Ground

A ground is a direct electrical connection to the earth, a connection to a particular point in an electrical or electronic circuit, or an indirect connection that operates as the result of capacitance between wireless equipment and the earth or a large mass of conductive material. The resistance of a grounding electrode is affected by the resistivity of the soil. Soil resistivity varies by the soil structure. The soil characteristics and the distribution of the current in the soil are very important in determining voltage effects on structures and the effectiveness of grounding and lightning protection systems.

Lightning Protection

Lightning effects cause thousands of personal injuries and deaths as well as billions of dollars in property loss annually. Damage prevention can be achieved through a properly designed and installed lightning protection mitigation system. Lightning protection systems are special grounding systems designed to safely conduct the extremely high voltage currents associated with lightning strikes.

System Care

To maintain grounding systems and equipment at an acceptable level of operational readiness, the following (minimum) action items should be performed:

Grounding Systems

- review maintenance records
- review operator records
- clean equipment
- perform equipment inspections
  - — conductors for damage
  - — connections for degradation
  - — insulation for discoloration
  - — loose grounding connections
- tighten grounding connections
- earth electrode measurements
- equipment ground measurements
- signal reference subsystem measurements

Lightning Protection

- review maintenance records
- review operator records
- perform equipment inspections
  - — connections for degradation
  - — structure surfaces for discoloration
  - — loose connections and components
  - — corrosion of connections and components
- tighten connections
- perform connection measurements
- recertify system (if required)

ILLUMINATION—BULBS AND BALLASTS

In the lighting world there are innumerable types and values of illuminating devices, bulbs, fixtures, components, systems and schemes. As a general rule, facility managers deal with three types, incandescent, fluorescent and gas discharge lamping. See Figure II-26.

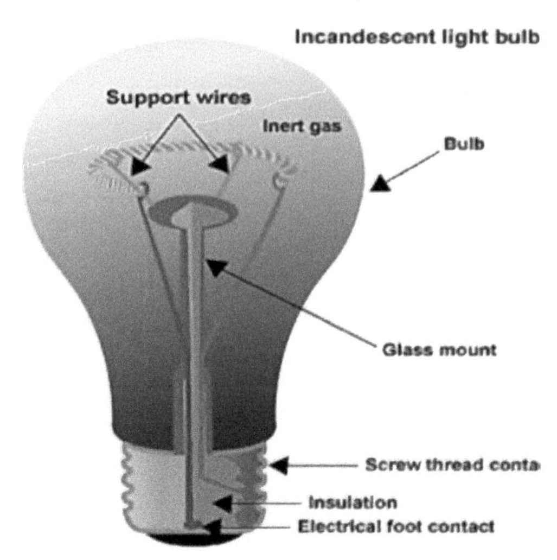

Figure II-26
Incandescent

In incandescent lamps, light is given off by a wire (filament) heated white hot. The amount of light that can be obtained from this type of lamp is limited by its physical dimensions and the manufacturing facilities. The important elements of the bulb are the tungsten filament, a mandrel on which the filament is wound and inert gasses which cool the filament to cut down on evaporation and carry any tungsten to the top of the lamp, creating a dark spot on its top after long use (evaporated tungsten). The gas is usually a combination of nitrogen and argon. To obtain more illumination from devices and still maintain practical dimensions, other types of lighting have been developed, such as fluorescent lamps, where brilliant lighting is required. These lamps contain a coating of fluorescent material applied to the inside of the glass tube that glows and gives off light when acted on by rays (similar to x-rays) produced by the flow of electricity. The fluorescent lamp unit generally consists of two electrodes, one at either end of the tube. The electrodes alternately emit and receive, and receive and emit electrons in a back and forth motion. A drop of mercury is added inside the tube to make it start or light up initially. A ballast is included in the circuit to permit the electrodes to reach sufficient voltage to start the arc going across the tube. An electrical ballast is a device intended to limit the amount of current in an electric circuit. A familiar and widely used example is the inductive ballast used in fluorescent lamps, to limit the current through the tube (which would otherwise rise to destructive levels due to the tube's negative resistance characteristic). Lamp starters are required on some inductor type ballasts. They connect both ends of the lamp to preheat the lamp ends for one second before lighting. See Figure II-27 and II-28.

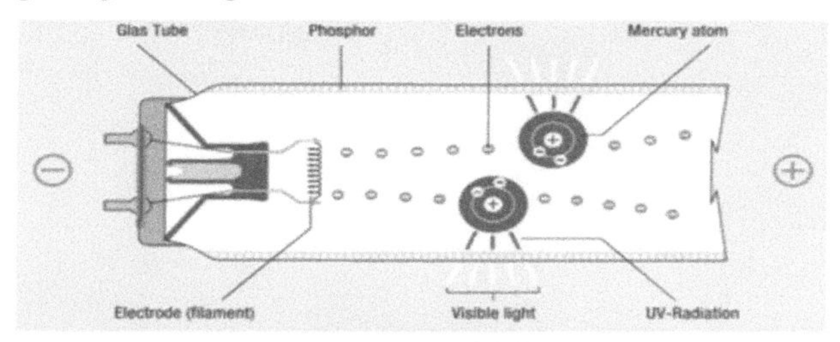

Fluorescent bulb

Starter

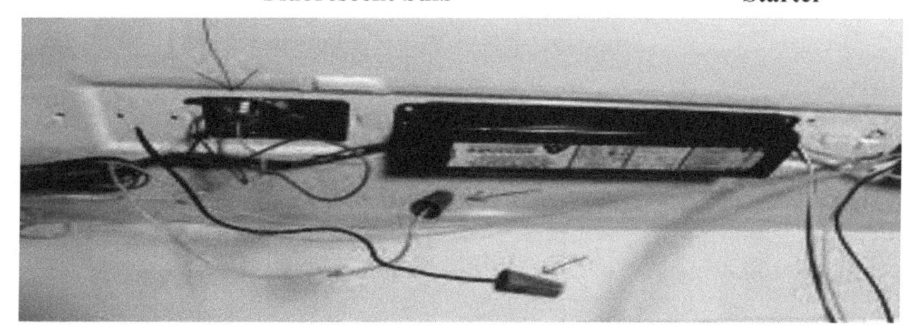

Typical Fuorescent Fixture Ballast Assembly

Figure II-27
Some Attributes of Fluorescents

Note: The days of the incandescent light bulb are waning. Soon it will be taken over by the CFL (compact fluorescent lamp), designed to replace the standard light bulb. It uses about four times less energy than that needed to power an equivalent incandescent lamp, operates much cooler, and lasts 8 to 15 times longer. The lamps use a tube which is curved or folded to fit into the space of an incandescent bulb, and a compact electronic ballast in the base of the lamp. Their principle of operation remains the same as in other fluorescent lighting.

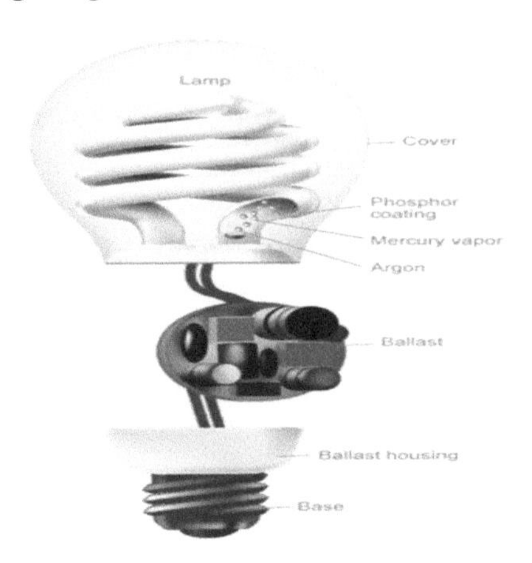

Figure II-28
Compact Fluorescent Lamp

Where high-intensity illumination is required (such as in parking lots) a gaseous discharge lamp fits the bill. The primary advantage of a gas discharge lamp is that it can give out much more light for its size than an incandescent lamp of comparable size. Sodium and mercury are commonly used materials for gas discharge lamps. With sodium, the principle of operation is the same, but the light produced is a highly intense yellow-orange. Gaseous discharge lamps are usually designed for multiple-circuit operation. Other materials are also used and are generally included in the class of halogen lamps.

In the mercury-vapor lamp, the mercury is vaporized by the current flowing through the lamp. (Again the current is controlled by a ballast). When the current continues through the gaseous mercury, the gas gives off great amounts of concentrated green-blue light. See Figure II-29.

MACHINE GUARDING

The purpose of machine guarding is to protect the machine operator and other employees in the work area from hazards created by in-going nip points, rotating parts, flying chips and sparks. Some examples of

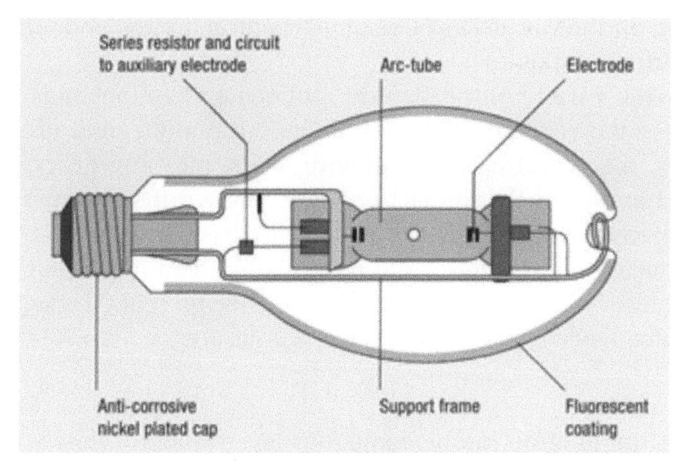

Figure II-29
Mercury Vapor Lamp

this are barrier guards, light curtains, two-hand operating devices, etc. [29 CFR 1910.212(a)(1)]/General Requirements: [29 CFR 1910.212(a)(2)]. See Figure II-30.

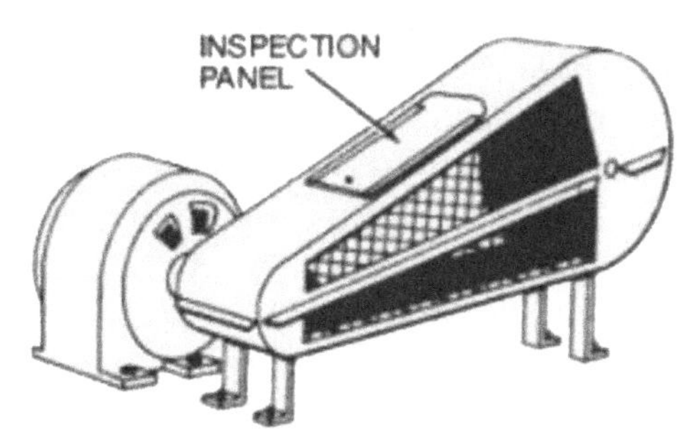

Figure II-30
Fixed Guard Enclosing a Belt and Pulleys

Moving machine parts have the potential to cause severe workplace injuries, such as crushed fingers or hands, amputations, burns or blindness. Safeguards are essential for protecting workers from these preventable injuries. Any machine part, function or process that may cause injury must be safeguarded. When the operation of a machine or accidental contact injures the operator or others in the vicinity, the hazards must be eliminated or controlled. This page contains general information on the various hazards of mechanical motion and techniques for protecting workers.

A wide variety of mechanical motions and actions may present hazards to the worker. These can include the movement of rotating members, reciprocating arms, moving belts, meshing gears, cutting teeth and any parts that impact or shear. These different types of hazardous mechanical motions and actions are basic in varying combinations to nearly all machines, and recognizing them is the first step toward protecting workers from the dangers they present. The basic types of hazardous mechanical motions and actions are:

Rotating Motion

Rotating motion can be dangerous; even smooth, slowly rotating shafts can grip hair and clothing, and through minor contact force the hand and arm into a dangerous position. Injuries due to contact with rotating parts can be severe. Collars, couplings, cams, clutches, flywheels, shaft ends, spindles, meshing gears and horizontal or vertical shafting are some examples of common rotating mechanisms which may be hazardous. The danger increases when projections such as set screws, bolts, nicks, abrasions, and projecting keys or set screws are exposed on rotating parts. In-running nip point hazards are caused by the rotating parts on machinery. There are three main types of in-running nips. Parts can rotate in opposite directions while their axes are parallel to each other. These parts may be in contact (producing a nip point) or in close proximity. In the latter case, stock fed between two rolls produces a nip point. As seen here, this danger is common on machines with intermeshing gears, rolling mills and calendars. Nip points are also created between rotating and tangentially moving parts. Some examples would be: the point of contact between a power transmission belt and its pulley, a chain and a sprocket, and a rack and pinion. Nip points can occur between rotating and fixed parts which create a shearing, crushing or abrading action. Examples are: spoke hand wheels or flywheels, screw conveyors, or the periphery of an abrasive wheel and an incorrectly adjusted work rest and tongue.

Reciprocating Motions

Reciprocating motions may be hazardous because, during the back-and-forth or up-and-down motion, a worker may be struck by or caught between a moving and a stationary part.

Transverse Motion

Transverse motion (movement in a straight, continuous line) creates a hazard because a worker may be struck or caught in a pinch or shear point by the moving part.

Controlling the Risk

Determine the appropriate control measures that must be put in place to eliminate the risk, or where it is not reasonably practicable to do so, the risk must be minimized. The hierarchy of control (listed in order of priority) is:

Elimination

Elimination (means to completely remove the hazard, or the risk of hazard exposure). Removal of the hazard is the ideal control solution.

Substitution

Substitution (involves replacing a hazardous piece of machinery or a work process with a non-hazardous one).

Engineering

If a hazard cannot be eliminated or replaced with a less hazardous option, the next preferred measure is to use an engineering control. Engineering controls may include machine guarding.

Figure II-31
Mandatory Compliance

Administration

Where engineering cannot fully control a health and safety risk, administration controls should be used. Administration controls introduce work practices that reduce risk and limit employee exposure. They include:

- training employees in correct and safe operation
- developing safe operating procedures (SOPs)
- reducing the number of employees exposed to the hazard
- reducing the period of employee exposure
- developing and implementing lock-out/tag-out/block-out procedures
- displaying appropriate warning signs

Machine hazards which may be controlled by guarding include:
- points of operation
- contact or entanglement with machinery
- being trapped between machine and material or fixed structure
- contact with material in motion
- being struck by ejected parts of machinery
- being struck by material ejected from machine
- release of potential energy

Making Sure Your Machine Guards Are in Place

Managing health and safety is an ongoing process that requires commitment by both management and employees in order to:
- identify the hazard
- assess the risk
- control the risk
- evaluate control measures

Safeguards Must Meet These Minimum General Requirements:

Prevent Contact

The safeguard must prevent hands, arms, and any other part of a worker's body from making contact with dangerous moving parts. A good safeguarding system eliminates the possibility of the operator or another worker placing parts of their bodies near hazardous moving parts.

Secure

Workers should not be able to easily remove or tamper with the safeguard, because a safeguard that can easily be made ineffective is no safeguard at all. Guards and safety devices should be made of durable material that will withstand the conditions of normal use. They must be firmly secured to the machine.

Protect from Falling Objects

The safeguard should ensure that no objects can fall into moving parts. A small tool which is dropped into a cycling machine could easily become a projectile that could strike and injure someone.

Create No New Hazards:

A safeguard defeats its own purpose if it creates a hazard of its own such as a shear point, a jagged edge, or an unfinished surface which can cause a laceration. The edges of guards, for instance, should be rolled or bolted in such a way that they eliminate sharp edges.

Create No Interference

Any safeguard which impedes a worker from performing the job quickly and comfortably might soon be overridden or disregarded. Proper safeguarding can actually enhance efficiency as it can relieve the worker's apprehensions about injury.

Allow Safe Lubrication

If possible, one should be able to lubricate the machine without removing the safeguards. Locating oil reservoirs outside the guard, with a line leading to the lubrication point, will reduce the need for the operator or maintenance worker to enter the hazardous area.

Maintenance and Repair Procedures

Good maintenance and repair procedures contribute significantly to the safety of the maintenance crew as well as that of machine operators. Workers should consider:

- activities they perform
- where their face, hands and feet are placed
- the body position they assume while performing a specific task
- hazard exposure inherent in the equipment or generated by it
- regular workplace inspections (using a checklist) to help uncover obvious workplace hazard

The variety and complexity of machines to be serviced, the hazards associated with their power sources, the special dangers that may be present during machine breakdown, and the severe time constraints often placed on maintenance personnel all make safe maintenance and repair work difficult. If possible, machine design should permit routine lubrication and adjustment without removal of safeguards. But when safeguards must be removed, and the machine serviced, the lockout procedure of 29 CFR 1910.147 must be adhered to. The maintenance and repair crew must never fail to replace the guards before the job is considered finished and the machine released from lockout. To prevent hazards while servicing machines, each machine or piece of equipment should be safeguarded during servicing or maintenance by:

- Notifying all affected employees (usually machine or equipment operators or users) that the machine or equipment must be shut down to perform some maintenance or servicing.
- Stopping the machine.
- Isolating the machine or piece of equipment from its energy source.
- Locking out or tagging out the energy source.
- Relieving any stored or residual energy.
- Verifying that the machine or equipment is isolated from the energy source.

Although this is the general rule, there are exceptions when the servicing or maintenance is not hazardous to an employee, when the servicing which is conducted is minor in nature, done as an integral part of production, and the employer utilizes alternative safeguards which provide effective protection as is required by 29 CFR 1910.212 or other specific standards. When the servicing or maintenance is completed, there are specific steps which must be taken to return the machine or piece of equipment to service. These steps include:

- Inspection of the machine or equipment to ensure that all guards and other safety devices are in place and functional.
- Checking the area to ensure that energizing and startup of the machine or equipment will not endanger employees.
- Removal of the lockout devices.
- Re-energizing of the machine or equipment.
- Notification of affected employees that the machine or equipment may be returned to service.

If it is necessary to oil machine parts while the machine is running, special safeguarding equipment may be needed solely to protect the oiler from exposure to hazardous moving parts. Maintenance personnel must know which machines can be serviced while running and which cannot. The danger of accident or injury is greatly reduced by shutting off and locking out all sources of energy.

PIPE HANGERS AND SUPPORTS

All pipe runs, whether vertical or horizontal, must be supported by an external means at regular intervals to prevent them from sagging due to their intrinsic self-weight and the weight of the fluids inside the pipe. Pipelines not supported at regular intervals will lead to sagging, causing stress at welded joints, flanges, etc., which may end in leaks and ultimately lead to failure or rupture in the line. Therefore it is essential to have a well designed pipe layout and pipe supports/hangers for long trouble-free service in the plant. See Figure II-32.

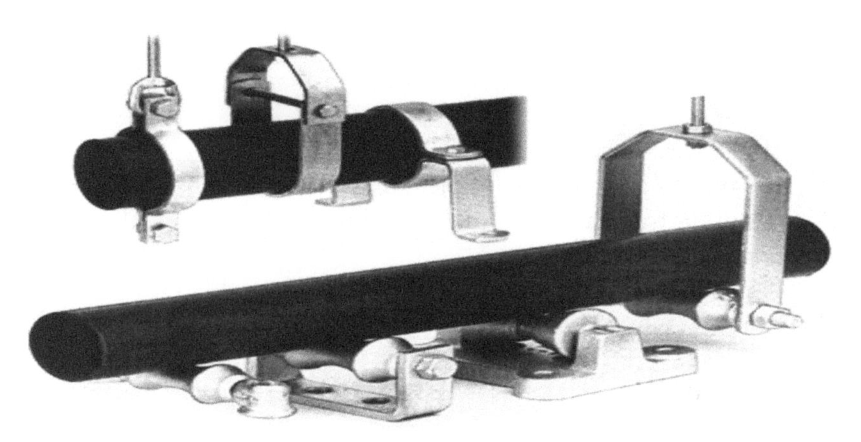

Figure II-32
Pipes Showing Supports

The design configuration of a pipe support assembly is dependent on the loading and operating conditions. The building material and type of pipe will determine the kind of support to use. The pipe run's length, location, and joints will determine the kind of support to use and where to use it (spacing). Regardless of material construction, all pipes have a self-weight based on diameter of pipe, wall thickness and specific gravity of the raw material used for manufacturing the pipe. This is expressed in pipe handbooks as weight per running meter. Specifications are compiled from generally accepted engineering principles and standards from the American National Standards Institute (ANSI), Manufacturers Standardization Society (MSS), Factory Mutual (FM), Underwriters Laboratories (UL) and other pertinent documents. Appropriate materials and protective coatings must be used to prevent failure from environmental and galvanic corrosion. Material that comes in contact with pipe shall be compatible with piping material so that neither has a deteriorating effect on the other.

Load Calculations

Hangers, supports, anchors and restraints must be selected to withstand all static and dynamic loading conditions which act upon the piping system and associated equipment. Piping supports and equipment must be considered as a total system and appropriate balance calculations made to determine load forces at critical stress points. Loading conditions to be considered may include but are not limited to:

- The total load of pipe, fittings, valves, insulation and any expected contents of the pipe.
- Thermal expansion and contraction.
- Stress from cycling of equipment or process.
- Vibration transmitted to or from equipment or terminal connection.
- Wind, snow or ice loading on outdoor piping.
- Loading due to seismic forces if required by code or specification.

Support Methods

Normally pipes are placed on pipe racks. They are supported either from the bottom or hung from the supporting structure, depending on layout and availability of supporting locations and structure. Pipe supports and pipe hangers are normally used to support pipelines by suspending them from structural members or supporting pipelines from the bottom. See Figure II-33.

Pipelines which carry fluids have a tendency to expand with increases in temperature. Metal will expand with rises in temperature. The pipeline will expand in a linear direction (when hot fluids are transported through it). Such pipelines can become very complex; consisting of horizontal runs, vertical risers, U bends, elbows, loops, etc., replete

Vertical Support Method

Vertical pipe must be supported to keep it in a straight; vertical position. The method of support is determined by the location of the pipe run and joists.

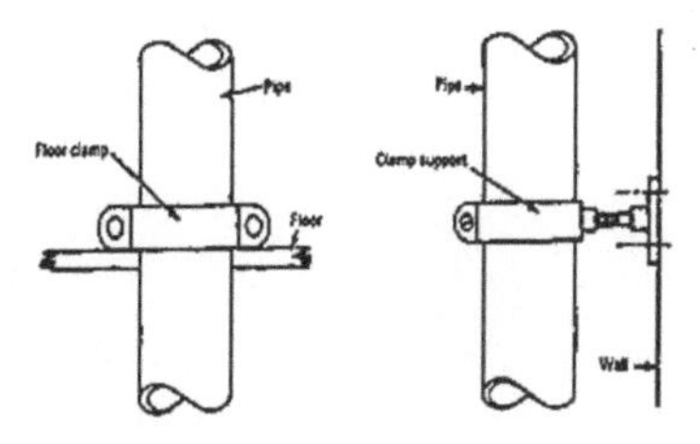

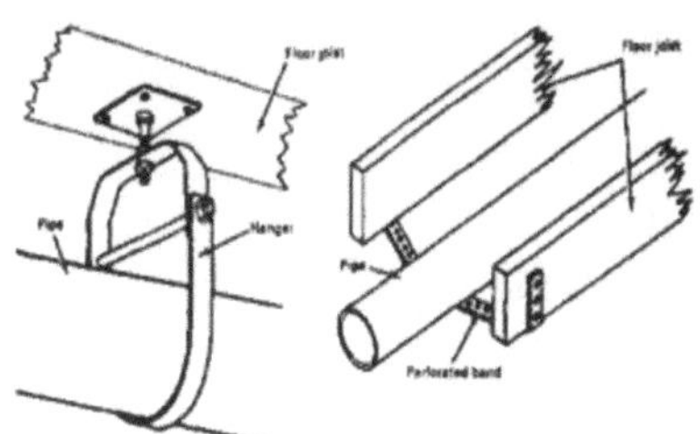

Horizontal Support Method

Horizontal pipes must be supported To keep them in line and prevent sagging. The method and location of support is determined by the type of pipe and location of the pipe runs.

Figure II-33
Support Methods

with fittings such as valves, flanges and insulation. The four main functions of a pipe support are to guide, anchor, absorb shock and support load. See Figure II-34.

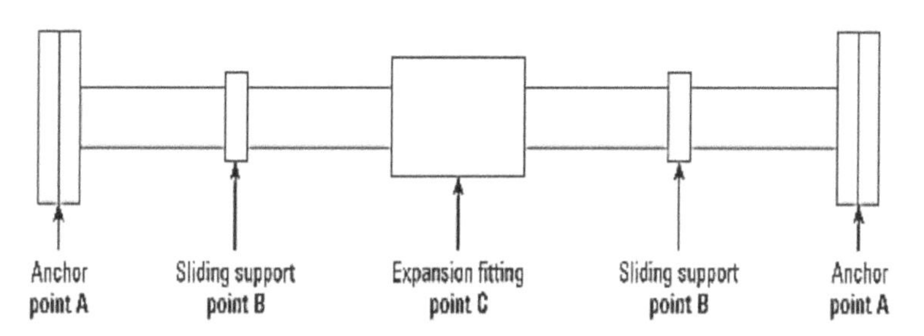

Figure II-34
Piping Support Points

An insulated pipe support (also called pre-insulated pipe support) is a load-bearing member and minimizes energy dissipation. Insulated pipe supports can be designed for vertical, axial and/or lateral loading combinations in both low- and high-temperature applications. Adequately insulating the pipeline increases the efficiency of the piping system by not allowing the "cold" or "hot" inside to escape to the environment.

Figure II-35
Prefabricated Insulated Supports

Supports Standards

The code ASME B 31.3 specifies under clause 321.1.1 the layout and design of piping and its supporting elements shall be directed toward preventing the following:

- Piping stresses in excess of those permitted in the Code.
- Leakage at joints.
- Excessive thrusts and moments on connected equipment (such as pumps and turbines).
- Excessive stresses in the supporting (or restraining) elements.
- Resonance with imposed or fluid-induced vibrations.
- Excessive interference with thermal expansion and contraction in piping which is otherwise adequately flexible.
- Unintentional disengagement of piping from its supports.
- Excessive piping sag in piping requiring drainage slope.
- Excessive distortion or sag of piping subject to creep under conditions of repeated thermal cycling.
- Excessive heat flow, exposing supporting elements to temperature extremes outside their design limits.

Types

Pipe hangers are manufactured in wide varieties; all having specific uses. Figure II-36 shows images of commonly used types of pipe hangers and supports identified by their names.

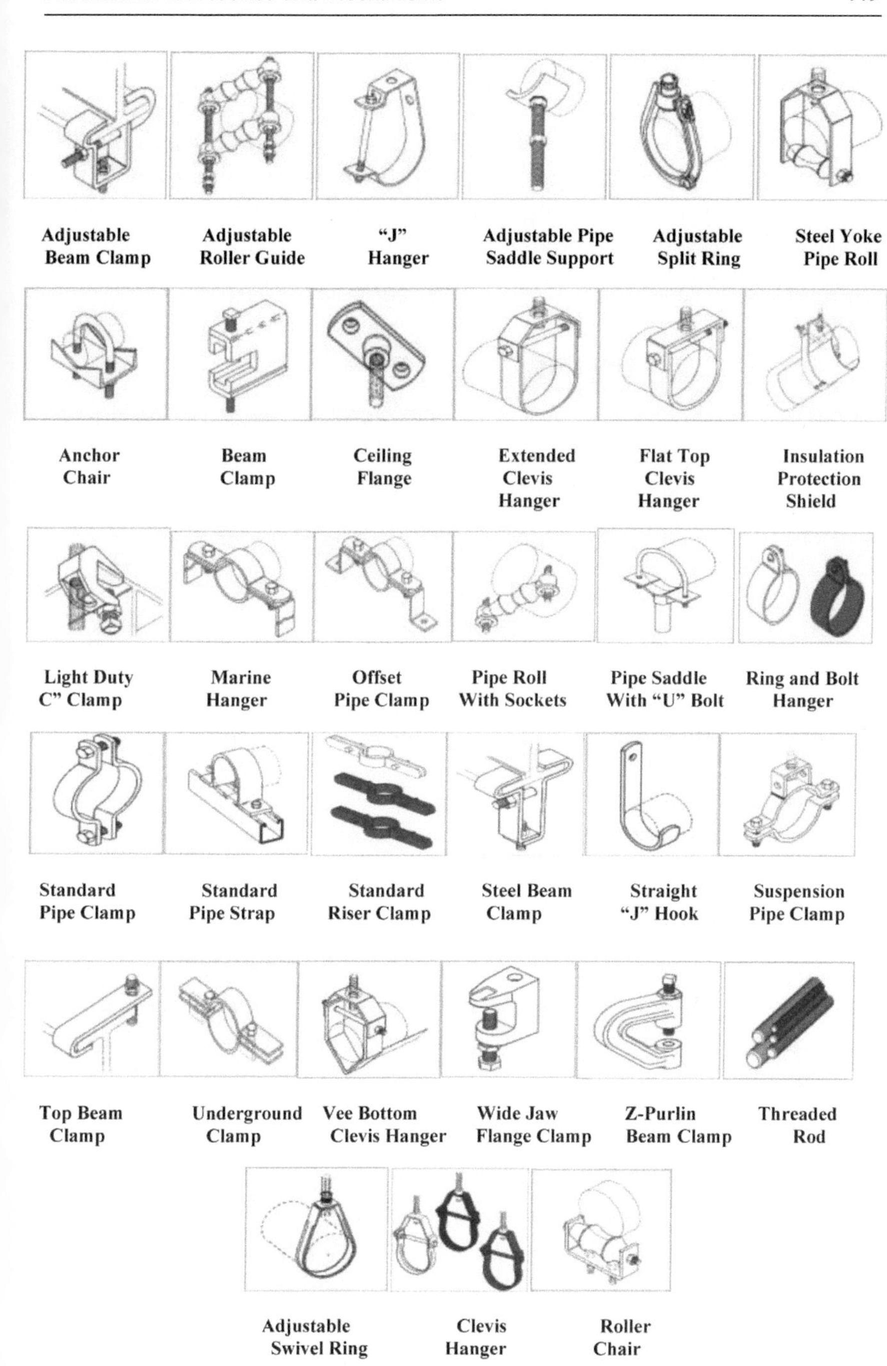

Figure II-36
Types of Pipe Hangers

PIPE MARKING/CONTENTS AND FLOW

ANSI/ASME A13.1 is the most common pipe identification standard used in the United States, and until the latest revision dated 2007, the standard has been unchanged for nearly half a century. The standard specifies the primary and secondary means of identifying pipe contents, as well as the size, color and placement of the identification device. See Figure II-37.

Figure II-37
Pipes Labeled for Content and Direction of Flow

Primary Identification

The legend (name of pipe content) and directional flow arrow remain the primary means of identifying pipe content. The size and placement of the marker arrow has not changed.

Secondary Identification

The secondary means of identification is the color code of the marker. That portion of the standard has changed dramatically. In addition, the terminology of inherently hazardous or non-hazardous has been removed from the standards. The combination of Yellow/Black is now assigned with flammable fluids, and Green/White shall now identify potable, cooling, boiler feed and other waters. These two changes mean that legends such as hot water, cold water and steam will now all use the color code of Green/White. The other significant color changes included the addition of Brown/White for combustible fluids and Orange/Black for toxic or corrosive fluids. The fact that the standard has identified specific colors for flammable fluids, combustible fluids and toxic or corrosive fluids means you must consult material safety data sheets before selecting a color. Further, if the pipe content contains multiple hazards (flammable and toxic) it must be determined which poses the greater risk and marked accordingly. For example, if chilled water or heating systems contain toxic treatments, the color combination should be Orange/Black. The new 2007 standard also identifies, for the first time, four additional color combinations and specifically identifies all of the exact background colors to be used. The exact colors are safety colors contained in the ANSI Z535.1-2007 standard. See Figure II-38.

Installation Guide

Visibility markers shall be located so that they are readily visible to plant personnel from the point of normal approach. Pipe markers instantly tell you all you need to know about pipe contents, direction of flow and whether hazardous or safe. See Figure II-39.

Sizing Recommendations

The A13.1-2007 standard also makes recommendations as to the size of letter height and length of color field for various pipe diameters. These recommendations are shown in the table. Markers, used properly with arrows and banding tape or arrow tape, will meet the standard. See Figure II-40.

Labeling

- Obtain a legend list of all pipe contents in your plant.
- Collect the following data on your piping systems. (This may require tracing lines to determine quantities and sizes).

PNEUMATIC CONTROLS

Pneumatic systems are mechanical systems that use compressed air for power. A pneumatic control is a control circuit that uses air pressure in conjunction with a mechanical system to perform regulating and control functions, such as increasing and decreasing temperature and other process variables such as pressure, flow and liquid level in large heating, ventilation and air-conditioning (HVAC) systems. Relying on sensors and thermostats to reduce or maintain line pressure from the sensor to the control device, compressed air is carried from a controller to a control device.

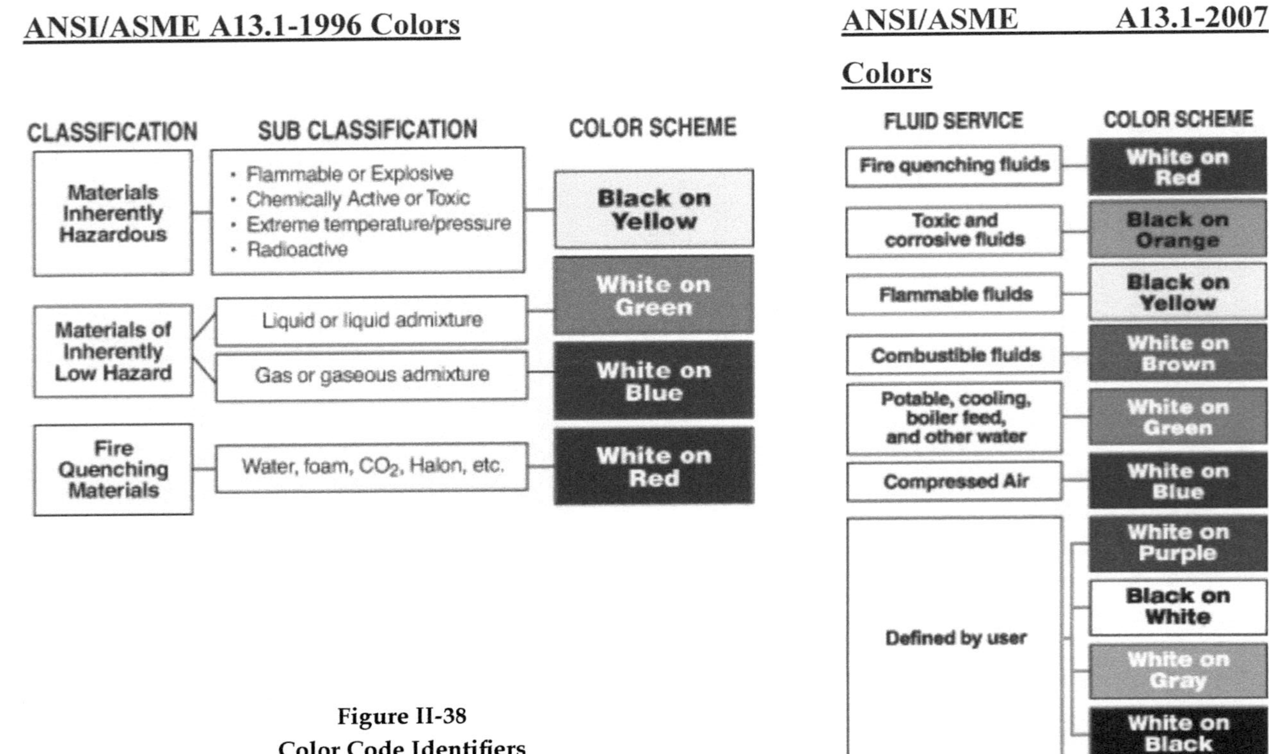

Figure II-38
Color Code Identifiers

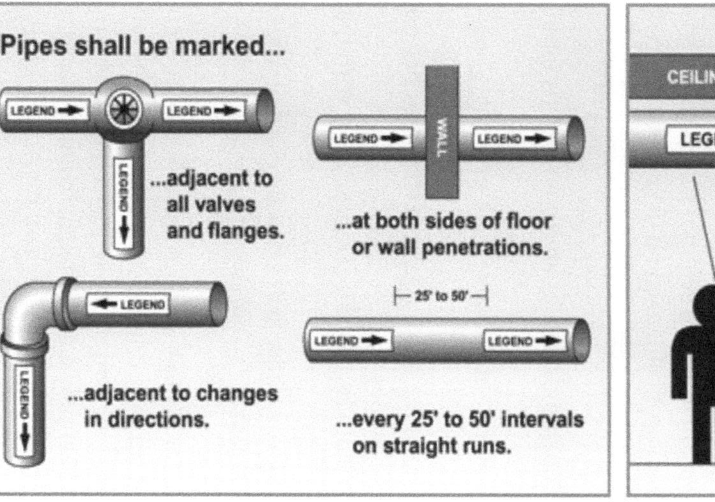

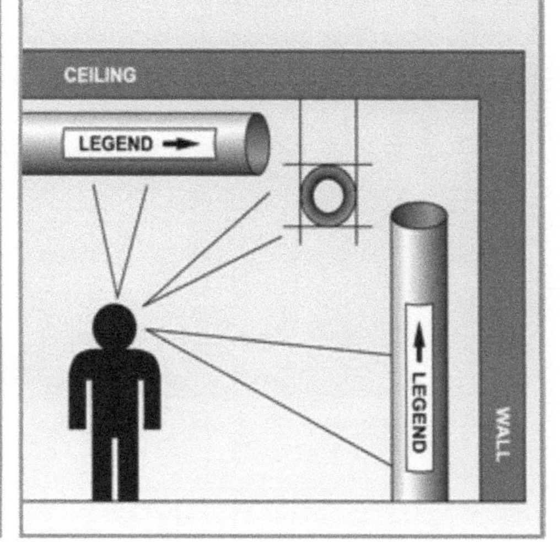

ABOVE: Figure II-39
Marking Locations

RIGHT: Figure II-40
Label Sizes

Fits Pipe Outer Diameter	Length Color Field	Letter Height
.75" - 1.25" (19mm - 32mm)	8" (203mm)	.5" (13mm)
1.5" - 2" (38mm - 51mm)	8" (203mm)	.75" (19mm)
2.5" - 6" (64mm - 152mm)	12" (305mm)	1.25" (32mm)
8" - 10" (203mm - 254mm)	24" (610mm)	2.50" (64mm)
over 10" (over 254mm)	32" (813mm)	3.50" (89mm)

NOTE: For pipes less than 3/4" in diameter, a permanently legible tag is recommended.

The controller responds to changes in temperature, humidity and static pressure, providing feedback that causes the actuator to open or close to meet the control setpoint. Pneumatic controls are often used in places where it is too dangerous to use electrical devices or where the air is too contaminated for electronics to work. See Figure II-41.

> **Note**: A calorifier is a heat exchanger which heats water indirectly by circulating it over one or more heating coils. In this case, steam (the heat source) is contained in a pipe immersed in the water of the heat exchanger vessel.

Controller Types

Continuous proportional control, as opposed to on-off control (home thermostats), throttles a control valve proportionately somewhere between fully on and fully off instead of only on or off. Proportional control allows sustained precise control right at a desired setpoint instead of the hunting or oscillation around a setpoint that is characteristic of on-off systems. Pneumatic controllers come in five basic types: "and," "or," "not," "memory" and "time-on/time-off," as follow:

- The AND controller will open the output valve if both of its input valves are pressurized.
- The OR controller will open the output valve if either input is pressurized.
- The NOT controller will open the output valve if the input has no pressure and will close it if it does.
- The MEMORY controller will keep its last position, even with the air pressure removed.
- The TIME-ON/TIME-OFF controller will open or close a valve, after a certain time has elapsed.

> **Note**: Pneumatic controls for heating systems must have a supply of clean, dry compressed air. As control air passes through very small passages and orifices, dirt, oil or moisture in the air cannot be tolerated. Most systems employ a dedicated air compressor and tank with sophisticated systems for filtering and drying the air.

Components

Pneumatic systems use gauges and valves to carefully regulate pressurized airflow through their systems. All pneumatic systems use a

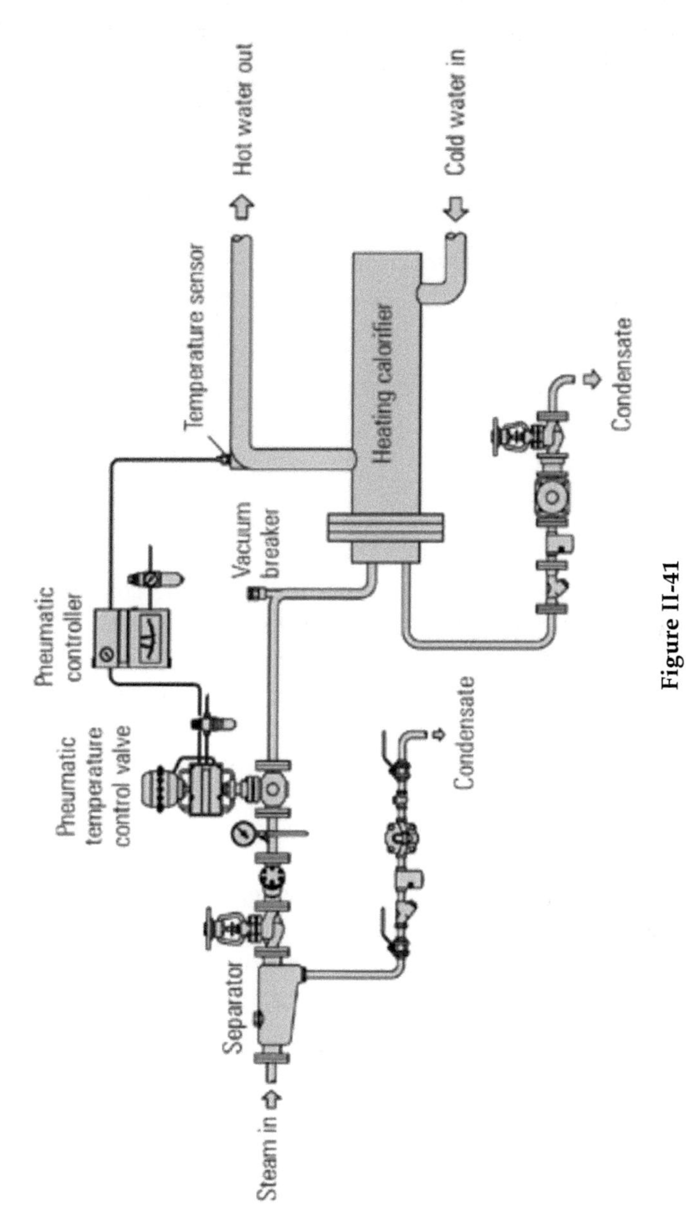

Figure II-41
Pneumatically Controlled Calorifier

source of compressed air to power moving parts, or actuators. Pneumatic systems commonly use compressed atmospheric air, as it is abundant and inexpensive. Pneumatic systems are quiet, cost-effective and easy to use. Typical system components follow.

Regulators and Gauges

Regulators and gauges are instruments attached to the compressor or compressor tank. The regulator is either mechanically or electrically triggered to release air into the pneumatic track. Gauges are mechanic or electric measuring instruments. They allow the operator or computer system to check and regulate the PSI of the air inside the compressor.

Check Valves

Check valves are one-way valves that are installed to the hose connecting the compressor or compressor tank to the buffer tank. They allow the compressed air to accumulate in the buffer tanks, but do not allow backflow into the compressor or compressor tank.

Buffer Tanks or Accumulators

Buffer tanks are secondary storage units for the compressed air originating from the compressor. They store the high-PSI compressed air for eventual use with the pneumatic actuators. These tanks help to prevent uneven airflow surges in the actuators, allow the compressor cycle to maximize its shutoff timing, and allow the compressor to be farther from the actuators in projects where this distance is useful or necessary.

Feed Lines

Feed lines are hoses or tubing that transfer pressurized air through the pneumatic system. The largest diameter hoses that can handle the PSI the system is using are installed. Large diameter hoses allow the pressurized air to travel quickly, eliminating airflow backups.

Directional Valves

Directional valves are placed before actuators. Multiple-valve systems are installed on projects with multiple actuators to power. Directional valves receive input from mechanical or electrical control sources. They redirect, stop or release the pressurized air to its appropriate actuators at the times desired. Directional valves can be triggered by the action of a button, spring, lever, pedal, solenoid or other device.

Actuators

An actuator is the component in a pneumatic system that does the work. There are numerous types of actuators, powered by pressurized air. Plunge and cylinder actuators are used frequently. The pressurized air is released into the cylinder to move a piston forward as the air is forced into the chamber. Pneumatic actuators are commonly used to actuate control valves and are available in two main forms, piston actuators and diaphragm actuators, as shown in Figure II-42.

Piston actuators are generally used where the stroke of a diaphragm actuator would be too short or the thrust is too small. The compressed air is applied to a solid piston contained within a solid cylinder. Piston actuators can be single acting or double acting, can withstand higher input pressures and can offer smaller cylinder volumes, which can act at high speed. See Figure II-42.

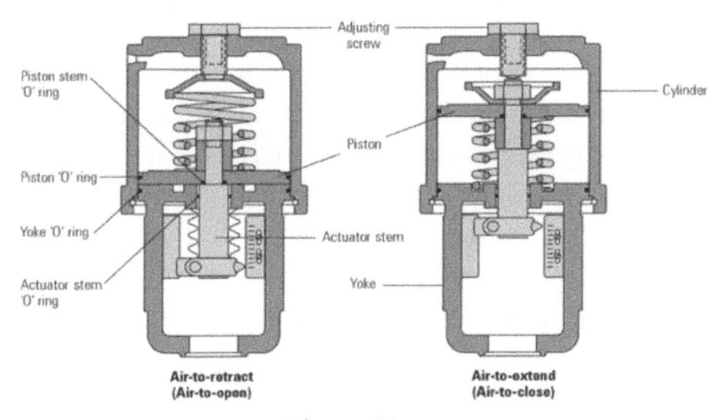

Figure II-42
Piston Actuators

Diaphragm actuators have compressed air supplied to a flexible membrane called the diaphragm. These types of actuators are single acting, in that air is only supplied to one side of the diaphragm, and they can be either direct acting (spring-to-retract) or reverse acting (spring-to-extend). See Figure II-43.

Functionality

Pneumatic controls for heating or HVAC systems typically consist of a pneumatic temperature sensor, a pneumatic controller and a valve. The sensor sends a signal to the controller which, in turn, controls the

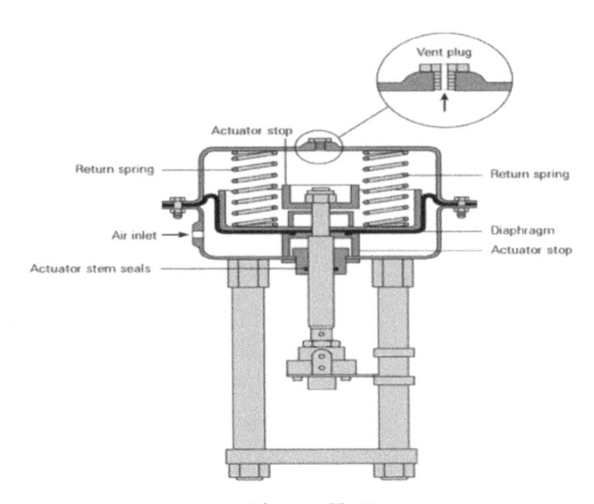

Figure II-43
Pneumatic Diaphragm Actuator

valve position. Process controllers compare a process variable such as a temperature from a temperature transmitter to a desired setpoint signal and move a proportional control valve such as a steam valve until the temperature equals the setpoint. In pneumatic control, a proportional 3 to 15 psi (pound per square inch) pneumatic signal communicates zero to 100 percent of full range. For a full temperature range of zero to 300°F, 3 psi would correspond to zero degrees, 15 psi would indicate 300°F, and 9 psi, or halfway between 3 and 15 psi, would indicate 150°F. Pneumatically driven control valves:

- can modulate large valves through tubing between the controller and the valve;
- are very accurate because pressure drop effects are nil due to leakage integrity.

Note: Pneumatic systems are very safe and pose no risk of fire or explosion. Since these systems use air from the environment, a leak will not cause contamination.

The positioner is attached to pneumatic control valves and is used to assign the valve stem's position to the electric input signal supplied by a control. It compares this signal to the travel or rotational angle of the control valve and produces the corresponding output signal pressure for the pneumatic actuator. The positioner mainly consists of an electric travel sensor system, an analog I/p module with a downstream booster as well as the electronics unit with a microcontroller. When a deviation occurs, the actuator is pressurized or vented. If required, the changes in the signal pressure can be slowed by a volume restriction. A constant air stream with a fixed setpoint to the atmosphere is created by a flow regulator with a fixed setpoint. The air stream is used to purge the inside of the case as well as to optimize the air capacity booster. The I/p module is supplied with a constant upstream pressure by the pressure-reducing valve to make it independent of the supply air pressure. See Figure II-44.

Figure II-44
Electro-pneumatic Positioner

Note: Pneumatic-control training topics include thermostat calibration, receiver-control calibration, relays, control adjustments, boilers and chillers, air conditioning units, steam converters and heat exchangers.

PRESSURE RELIEF VALVES

All pressurized systems require safety devices to protect people, processes and property. The principle types of devices used to prevent overpressure in plant settings are pressure relief valves (for air, steam and water service), rupture disks and fusible plugs.

Valve Types

Distinctions between types of these valves have been defined by the ASME (American Society of Mechanical Engineering). The ASME/ANSI PTC25.3 standards applicable to the USA define the following generic terms:

- *Pressure Relief Valve*

A spring-loaded pressure relief valve which is designed to open to relieve excess pressure and to reclose and prevent the further flow of fluid after normal conditions have been restored. It is characterized by a rapid-opening "pop" action or by opening in a manner generally proportional to the increase in pressure over the opening pressure. It may be used for either compressible or incompressible fluids, depending on design, adjustment, or application. This is a general term, which includes safety valves, relief valves and safety relief valves. See Figure II-45.

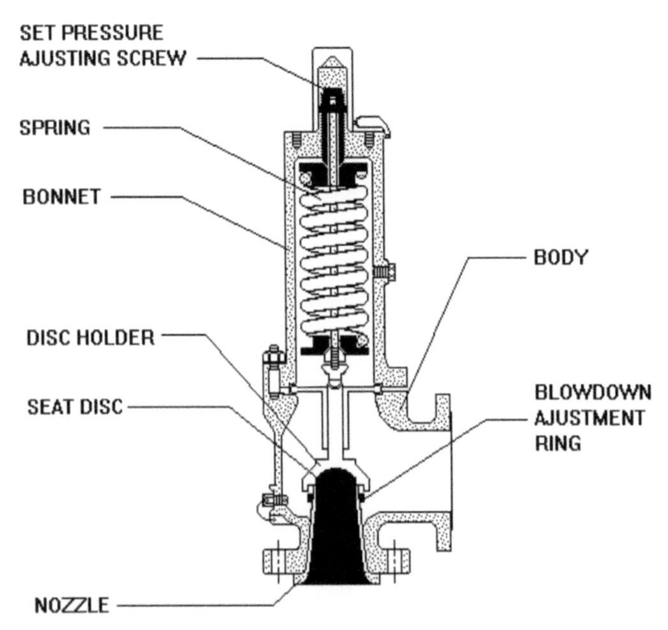

Figure II-45
Conventional Spring Operated Pressure Relief Valve

- *Safety Valve*

A pressure relief valve actuated by inlet static pressure and characterized by rapid opening or pop action. Safety valves are primarily used with compressible gases and in particular for steam and air services. However, they can also be used for process type applications where they may be needed to protect the plant or to prevent spoilage of the product being processed.

- *Relief Valve*

A pressure relief device actuated by inlet static pressure having a gradual lift generally proportional to the increase in pressure over opening pressure. Relief valves are commonly used in liquid systems, especially for lower capacities and thermal expansion duty. They can also be used on pumped systems as pressure overspill devices.

- *Safety Relief Valve*

A pressure relief valve characterized by rapid opening or pop action, or by opening in proportion to the increase in pressure over the opening pressure, depending on the application, and which may be used either for liquid or compressible fluid. In general, the safety relief valve will perform as a safety valve when used in a compressible gas system, but it will open in proportion to the overpressure when used in liquid systems, as would a relief valve.

"Pop" Type Valves

When the inlet static pressure rises above the set pressure of the safety valve, the disc will begin to lift off of its seat. However, as soon as the spring starts to compress, the spring force will increase. This means that the pressure would have to continue to rise before any further lift could occur, and for there to be any significant flow through the valve. The additional pressure rise required before the safety valve will discharge at its rated capacity is called the overpressure. The allowable overpressure depends on the standards being followed and the particular application. For compressible fluids, this is normally between 3% and 10% and for liquids between 10% and 25%. To achieve full opening from this small overpressure, the disc arrangement has to be specially designed to provide rapid opening. This is usually done by placing a shroud, skirt or hood around the disc. The volume contained within this shroud is known as the control or huddling chamber. See Figure II-46.

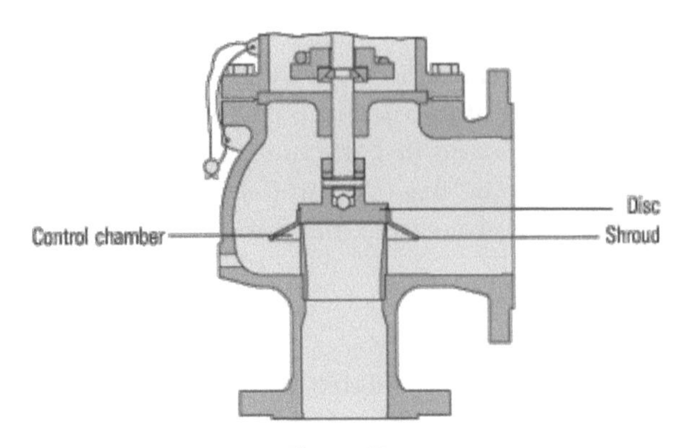

Figure II-46
Disc and Shroud Arrangement of a Rapid Opening Safety Valve

As lift begins and fluid enters the chamber, a larger area of the shroud is exposed to the fluid pressure. Because the magnitude of the lifting force (F) is proportional to the product of the pressure (P) and the area exposed to the fluid (A), (F = P x A), the opening force is increased. This incremental increase in opening force overcompensates for the increase in spring force, causing rapid opening. At the same time, the shroud reverses the direction of the flow, which provides a reaction force, further enhancing the lift.

Rupture Disks

Non-reclosing devices are those which are designed to remain open after operation. A manual means of resetting is usually provided.

- Bursting, or rupture discs, consist of a membrane or thin metal disk that will burst at a set pressure, relieving any overpressure. Although they can be used by themselves, on many applications they are used in conjunction with pressure relief valves. A rupture disc can be installed either on the inlet or outlet side of the valve. If installed on the inlet, it isolates the contained media from the valve. When there is an overpressure situation, the rupture disc bursts allowing the flow into the pressure relief valve, which will then lift. This arrangement is used to protect the internals of the valve from corrosive fluids. Rupture discs can also be installed alongside a safety valve as a secondary relief device. See Figure II-47.

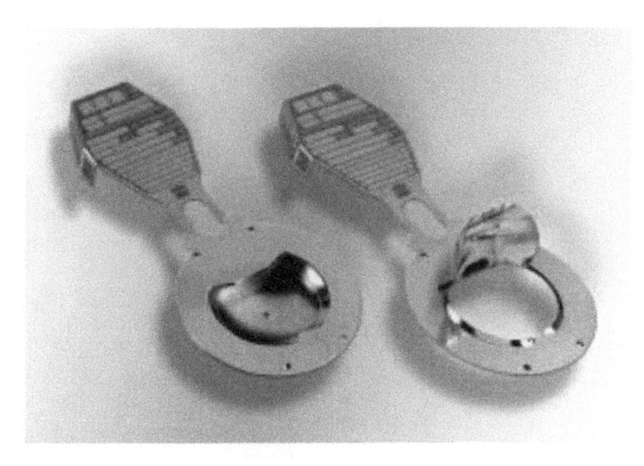

Figure II-47
New and Activated Rupture Disks

- Alternatively, if the safety valve discharges into a manifold containing corrosive media, a rupture disc can be installed on the safety valve outlet, preventing any flow from the manifold contacting the internals of the valve in normal use.

- Rupture discs are leak tight and low cost, but they require replacing after each operation. Most rupture disc installations contain a mechanism to indicate when the disc has ruptured and that it needs to be replaced.

Fusible Plugs

These consist of a plug with a lower melting point than the maximum operating temperature of the system that it is used to protect. In old steam locomotives, this type of device was used to dump the boiler water onto the fire if over-temperature occurred. See Figure II-48.

Pressure Terminology

Operating pressure (working pressure) is the gauge pressure existing at normal operating conditions within the system.

Set pressure is the gauge pressure at which, under operating conditions, direct loaded safety valves commence to lift.

Reseating pressure is the gauge pressure at which the direct loaded safety valve is re-closed.

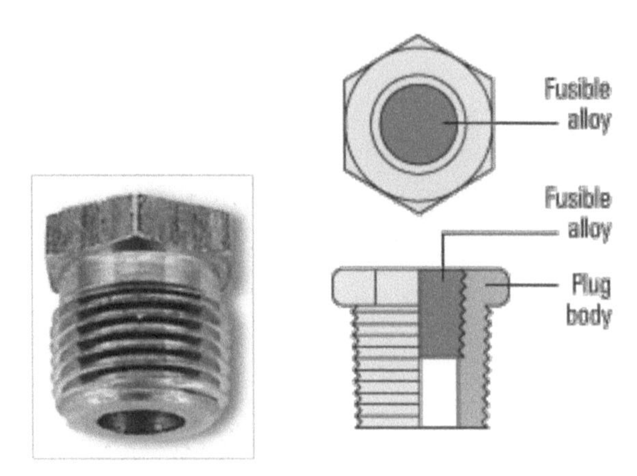

Figure II-48
Fusible Plug; Showing Melt Able Alloy

Accumulation is the increase in pressure over the maximum allowable working gauge pressure of the system to be protected.

Blowdown (reseating pressure difference) is the difference between actual popping pressure and actual reseating pressure, usually expressed as a percentage of set pressure or in pressure units.

Overpressure is a pressure increase over the set pressure of a pressure relief valve, usually expressed as a percentage of set pressure.

Maximum allowable working pressure (MAWP) is the maximum gauge pressure permissible at the top of a completed vessel in its operating position for a designated temperature.

Maximum allowable accumulated pressure (MAAP) is the maximum allowable working pressure plus the accumulation as established by reference to the applicable codes for operating or fire contingencies.

STEAM TRAPS

Steam traps are installed in steam lines to drain condensate from the lines without allowing the escape of steam. There are many designs of steam traps for high- and low-pressure steam service. Several kinds of thermostatic steam traps are used. In general, these traps are more compact than other types and have fewer moving parts than most mechanical steam traps. Some facilities may use continuous-flow steam traps of the orifice type in some constant service steam systems, oil-heating steam systems, ventilation pre-heaters, and other systems or services in which condensate forms at a fairly constant rate. Orifice-type steam traps are not suitable for services in which the condensate formation is not continuous. A steam trap consists of a valve and a device or arrangement that causes the valve to open and close as necessary to drain the condensate from piping without allowing the escape of steam. Steam traps are installed at low points in the system or machinery to be drained. Types of steam traps that are typically used in facilities are described below, with cutaway views.

- Ball Float Steam Trap
- Bucket Steam Trap
- Bellows Type Steam Trap
- Impulse Steam Trap

A control orifice runs through the disk from top to bottom; the disk is considerably smaller at the top than at the bottom. The bottom part of the disk extends through and beyond the orifice in the seat. The upper part of the disk (including the flange) is inside a cylinder. The cylinder tapers inward, so the amount of clearance between the flange and the cylinder varies according to the position of the valve. When the valve is open, the clearance is greater than when the valve is closed.

Steam Trap Types

Ball Float Steam Trap

As illustrated, the valve of this trap is connected to the float in such a way that the valve opens when the float rises. When the trap is in operation, the steam and any water that may be mixed with it flows into the float chamber. Being heavier than the steam, the water falls to the bottom of the trap, causing the water level to rise. As the water level rises, it lifts the float, thus lifting the valve plug and opening the valve. The condensate drains out and the float moves down to a lower position, closing the valve before the condensate level gets low enough to allow steam to escape. The condensate that passes out of the trap is returned to the feed system. See Figure II-49.

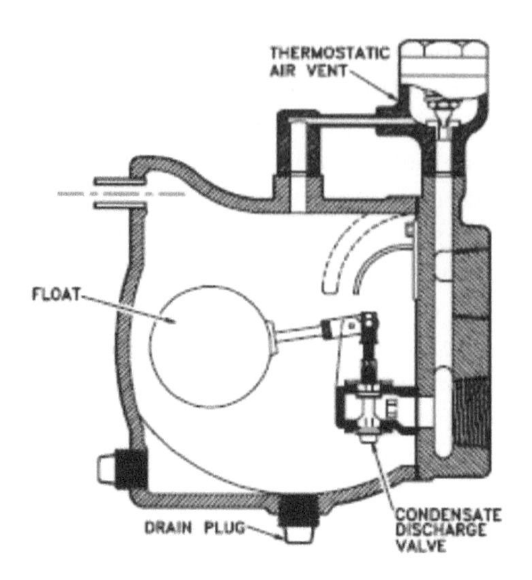

Figure II-49
Thermostatic Ball Valve Trap

Bucket Trap

As condensate enters the trap body, the bucket floats. The valve is connected to the bucket in such a way that the valve closes as the bucket rises. As condensate continues to flow into the trap body, the valve remains closed until the bucket is full. When the bucket is full, it sinks and thus opens the valve. The valve remains open until condensate has passed out to allow the bucket to float, closing the valve. See Figure II-50.

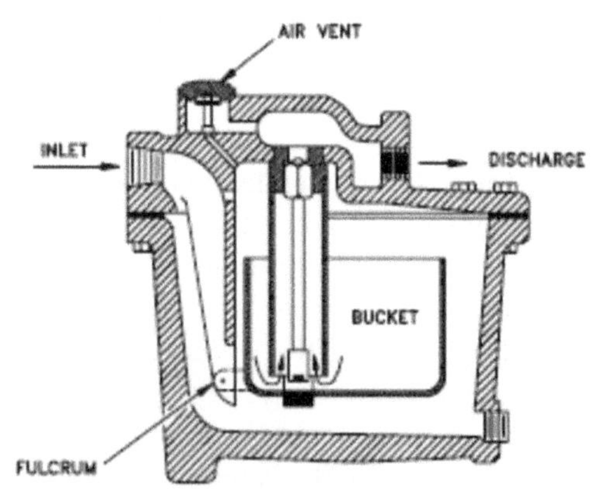

Figure II-50
Bucket Trap

Bellows-types Steam Trap

A bellows-type steam trap is shown in Figure II-51. The operation of this trap is controlled by the expansion of the vapor of a volatile liquid which is enclosed in a bellows-type element. Steam enters the trap body and heats the volatile liquid in the sealed bellows, causing expansion of the bellows. The valve is attached to the bellows in such a way that the valve closes when the bellows expands. The valve remains closed, trapping steam in the valve body. As the steam cools and condenses, the bellows cools and contracts, thereby opening the valve and allowing the condensate to drain.

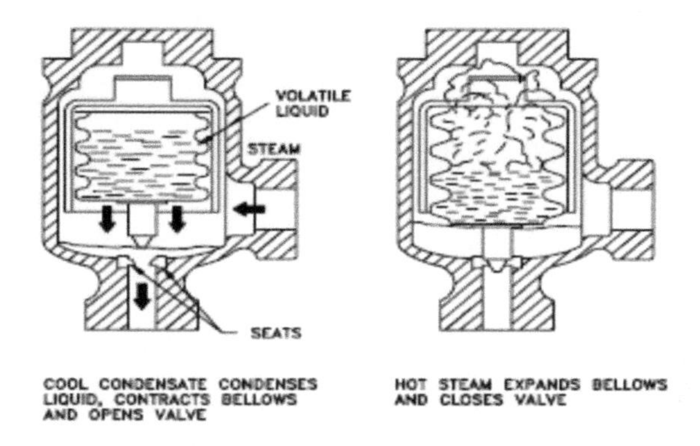

Figure II-51
Bellows Type Trap

Impulse Steam Trap

As shown in Figure II-52, impulse steam traps pass steam and condensate through a strainer before entering the trap. A circular baffle keeps the entering steam and condensate from impinging on the cylinder or on the disk. The impulse type of steam trap is dependent on the principle that hot water, under pressure, tends to flash into steam when the pressure is reduced. The only moving part in the steam trap is the disk. A flange near the top of the disk acts as a piston. As demonstrated, the working surface above the flange is larger than the working surface below the flange.

A control orifice runs through the disk from top to bottom, and the disk is considerably smaller at the top than at the bottom. The bottom part of the disk extends through and beyond the orifice in the seat. The upper part of the disk (including the flange) is inside a cylinder. The cylinder tapers inward, so the amount of clearance between the flange

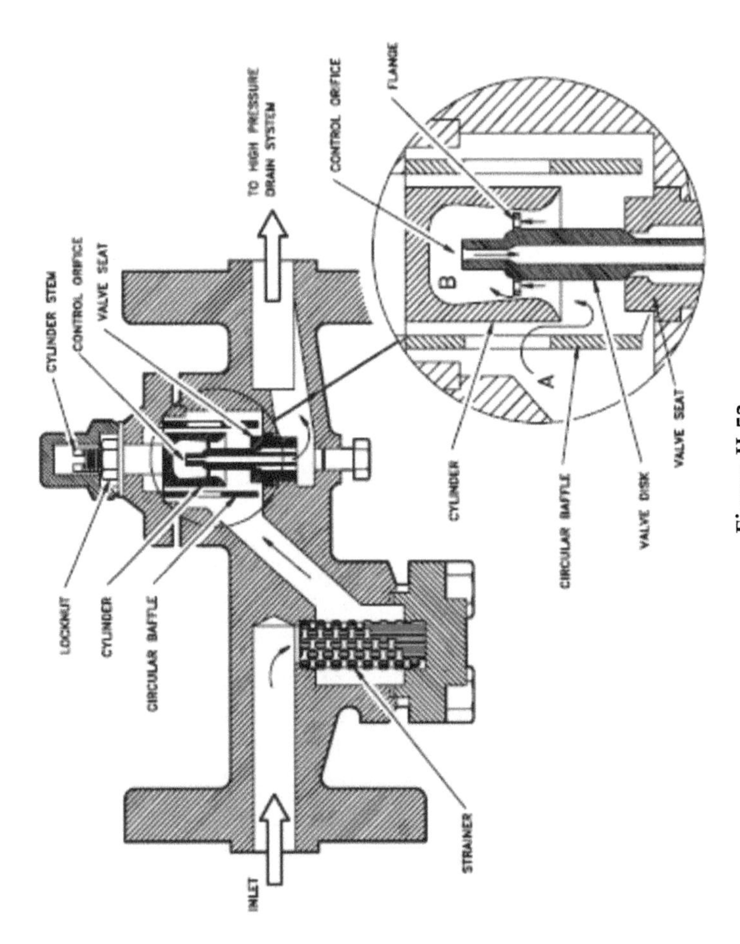

Figure II-52
Impulse Steam Trap

and the cylinder varies according to the position of the valve. When the valve is open, the clearance is greater than when the valve is closed.

Operation

When the trap is first placed in service, pressure from the inlet (chamber A) acts against the underside of the flange and lifts the disk off the valve seat. Condensate is thus allowed to pass out through the orifice in the seat. At the same time, a small amount of condensate (called control flow) flows up past the flange and into chamber B. The control flow discharges through the control orifice, into the outlet side of the trap, and the pressure in chamber B remains lower than the pressure in chamber A. As the line warms, the temperature of the condensate flowing through the trap increases.

The reverse taper of the cylinder varies the amount of flow around the flange until a balanced position is reached in which the total force exerted above the flange is equal to the total force exerted below the flange. It is important to note that there is still a pressure difference between chamber A and chamber B. The force is equalized because the effective area above the flange is larger than the effective area below the flange. The difference in working area is such that the valve maintains at an open, balanced position when the pressure in chamber B is approximately 86% of the pressure in chamber A.

As the temperature of the condensate approaches its boiling point, some of the control flow going to chamber B flashes into steam as it enters the low pressure area. Because the steam has a much greater volume than the water from which it is generated, pressure builds up in the space above the flange (chamber B). When the pressure in this space is 86% of the inlet pressure (chamber A), the force exerted on the top of the flange pushes the entire disk downward and closes the valve. With the valve closed, the only flow through the trap is past the flange and through the control orifice.

When the temperature of the condensate entering the trap drops slightly, condensate enters chamber B without flashing into steam. Pressure in chamber B is thus reduced to the point where the valve opens and allows condensate to flow through the orifice in the valve seat. The cycle is repeated continuously.

With a normal condensate load, the valve opens and closes at frequent intervals, discharging a small amount of condensate at each opening. With a heavy condensate load, the valve remains open and allows a continuous discharge of condensate. Although there are several variations of the orifice-type steam trap, they each have one thing in common—no moving parts. One or more restricted passageways or orifices allow condensate to trickle through, but do not allow steam to flow through. Some orifice-type steam traps have baffles in addition to orifices.

VALVE TYPES

Valve Construction

Valves are used to control fluids in closed systems and are typed or classified according to their use. They are usually made of bronze, brass, cast or malleable iron, or steel. Steel valves are either cast or forged and are made of either plain steel or alloy steel. Alloy steel valves are used in high-pressure, high-temperature systems; the disks and seats (internal sealing surfaces) of those valves are usually surfaced with a chromium cobalt alloy known as stellite. Stellite is extremely hard. Brass and bronze valves are never used in systems where temperatures exceed 550°F. Steel valves are used for all services above 550°F and in lower temperature systems where internal or external conditions of high pressure, vibration, or shock would be too severe for valves made of brass or bronze. Bronze valves are used almost exclusively in systems that carry salt water. The seats and disks of these valves are usually made of Monel, a metal that has excellent corrosion and erosion-resistant qualities.

Valve Types

Although many different types of valves are used to control the flow of fluids, the basic valve types can be divided into two general groups, stop valves and check valves. Besides the basic types of valves, many special valves, which cannot really be classified as either stop valves or check valves, are found in the facility's mechanical spaces. Many of those valves serve to control the pressure of fluids and are known as pressure-control valves. Other valves are identified by names that indicate their general function, such as thermostatic re-circulating valves.

Stop Valves

Stop valves are used to shut off or, in some cases, partially shut off the flow of fluid. Stop valves are controlled by the movement of the valve stem. Stop valves can be divided into four general categories: globe, gate, butterfly and ball valves. Plug valves and needle valves may also be considered stop valves.

Globe Valves

Globe valves are probably the most common valves in existence. The globe valve derives its name from the globular shape of the valve body. However, positive identification of a globe valve must be made internally because other valve types may have globular appearing bodies. Globe valve inlet and outlet openings are arranged in several ways to suit varying requirements of flow. Figure II-53 shows the common types of globe valve bodies: straight flow, angle flow and cross flow. Globe valves are used extensively throughout the engineering plant in a variety of systems.

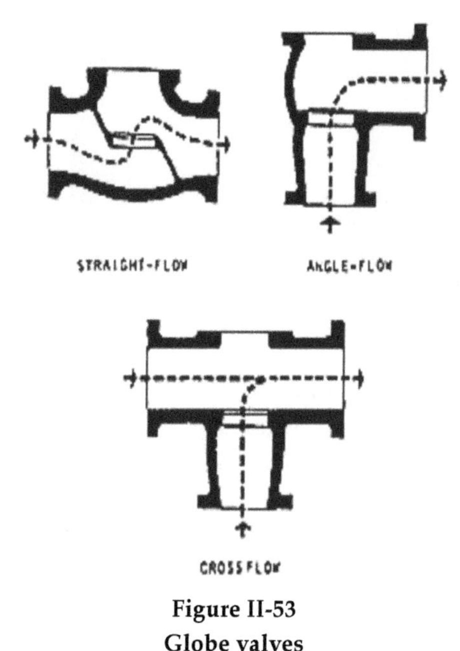

Figure II-53
Globe valves

Gate Valves

Gate valves are used when a straight-line flow of fluid and minimum restriction is desired. Gate valves are so named because the part that either stops or allows flow through the valve acts somewhat like the opening or closing of a gate and is called, appropriately, the gate. The gate is usually wedge shaped. When the valve is wide open, the gate is fully drawn up into the valve, leaving an opening for flow through the valve the same size as the pipe in which the valve is installed. Therefore, there is little pressure drop or flow restriction through the valve. Gate valves are not suitable for throttling purposes since the control of flow would be difficult due to valve design and since the flow of fluid slapping against a partially open gate can cause extensive damage to the valve. Gate valves are classified as either rising stem or non-rising stem valves.

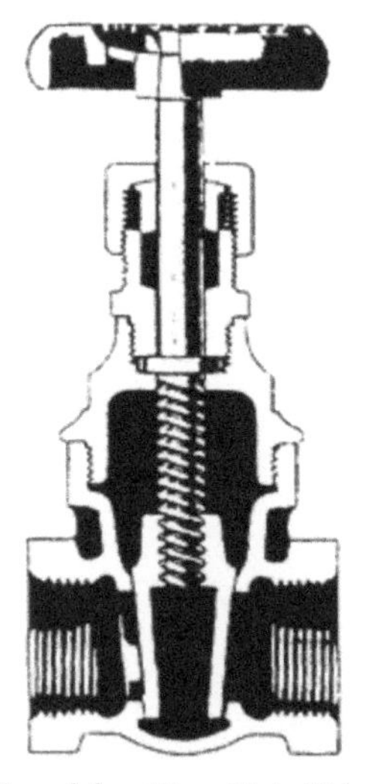

Non-rising Stem Gate Valve

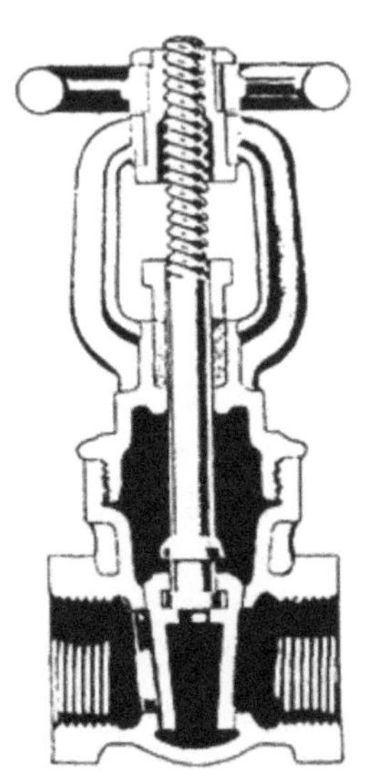

Rising Stem Gate Valve

Figure II-54
Gate Valves

On the non-rising-stem gate valve, the stem is threaded on the lower end into the gate. As the hand-wheel on the stem is rotated, the gate travels up or down the stem on the threads, while the stem remains vertically stationary. This type of valve is threaded onto the upper end of the stem to indicate valve position. The rising-stem gate valve has the stem attached to the gate; the gate and stem rise and lower together as the valve is operated. Gate valves used in steam systems have flexible gates. The reason for using a flexible gate is to prevent binding of the gate within the valve when the valve is in the closed position. When steam lines are heated, they will expand, causing some distortion of valve bodies. If a solid gate fits snugly between the seat of a valve in a cold steam system, when the system is heated and pipes elongate, the seats will compress against the gate, wedging the gate between them and clamping the valve shut. This problem is overcome by use of a flexible gate (two circular plates attached to each other with a flexible hub in the middle). This design allows the gate to flex as the valve seat compresses it, thereby preventing clamping.

Butterfly Valves

The butterfly valve, one type of which is shown in Figure II-55, may be used in a variety of systems. These valves can be used effectively in freshwater, saltwater, lube oil, and chill water systems. The butterfly valve is light in weight, relatively small, relatively quick acting, provides positive shutoff, and can be used for throttling. The butterfly valve has a body, a resilient seat, a butterfly disk, a stem, packing, a notched positioning plate, and a handle. The resilient seat is under compression when it is mounted in the valve body, thus making a seal around the periphery of the disk and both upper and lower points where the stem passes through the seat.

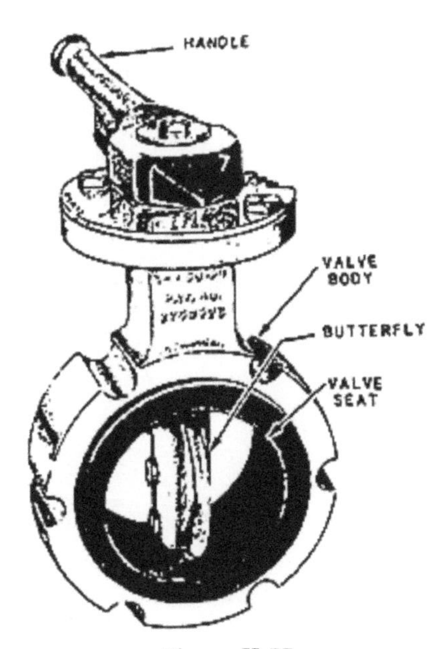

Figure II-55
Butterfly Valve

Packing is provided to form a positive seal around the stem for added protection in case the seal formed by the seat should become damaged. To close or open a butterfly valve, turn the handle only one quarter turn to rotate the disk 90°. Some larger butterfly valves may have a hand wheel that operates through a gearing arrangement to operate the valve. This method is used especially where space limitation prevents use of a long handle. Butterfly valves are relatively easy to maintain. The resilient seat is held in place by mechanical means, and neither bonding nor cementing is necessary. Because the seat is replaceable, the valve seat does not require lapping, grinding, or machine work.

Ball Valves

As the name implies, ball valves are stop valves that use a ball to stop or start the flow of fluid. The ball performs the same function as the disk in the globe valve. When the valve handle is operated to open the valve, the ball rotates to a point where the hole through the ball is in line with the valve body inlet and outlet. When the valve is shut, which requires only a 90-degree rotation of the hand wheel for most valves, the ball is rotated so the hole is perpendicular to the flow openings of the valve body, and flow is stopped. Most ball valves are of the quick-acting type (requiring only a 90-degree turn to operate the valve either completely open or closed), but many are planetary gear operated. This type of gearing allows the use of a relatively small hand wheel and operating force to operate a fairly large valve. The gearing does, however, increase the operating time for the valve. Some ball valves contain a swing check located within the ball to give the valve a check valve feature. See Figure II-56.

Check valves are used to allow fluid flow in a system in only one direction. They are operated by the flow of fluid in the piping. A check valve may be the

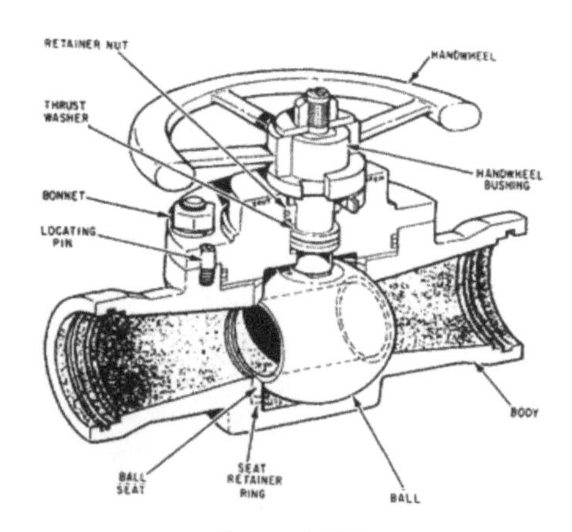

Figure II-56
Ball valve

swing type, lift type, or ball type. Most valves can be classified as being either stop valves or check valves. Some valves, however, function either as stop valves or as check valves, depending on the position of the valve stem. These valves are known as stop-check valves. A stop-check valve is shown in the cross section of Figure II-57. Stop-check valves are widely used throughout the physical plant, in drain lines and on the discharge side of many pumps. This type of valve looks very much like a lift check valve. However, the valve's stem is long enough so when it is screwed all the way down it holds the disk firmly against the seat, thus preventing any flow of fluid. In this position, the valve acts as a stop valve. When the stem is raised, the disk can be opened by pressure on the inlet side. In this position, the valve acts as a check valve, allowing the flow of fluid in only one direction. The maximum lift of the disk is controlled by the position of the valve stem. Therefore, the position of the valve stem limits the amount of fluid passing through the valve even when the valve is operating as a check valve.

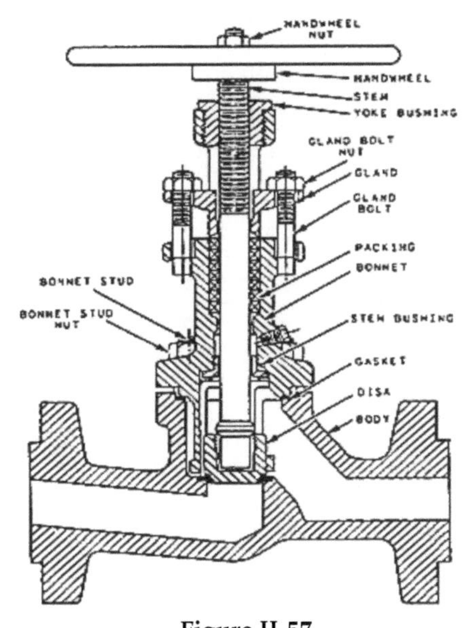

Figure II-57
Special Purpose Valves

Special Purpose Valves

There are many types of automatic pressure control valves. Some of them merely provide an escape for pressures exceeding the normal pressure; some provide only for the reduction of pressure; and some provide for the regulation of pressure.

Relief Valves

Relief valves are automatic valves used on system lines and equipment to prevent over pressurization. Most relief valves simply lift (open) at a preset pressure and reset (shut) when the pressure drops only slightly below the lifting pressure. Figure II-58 shows a relief valve of this type. System pressure simply acts under the valve disk at the inlet of the valve. When system pressure exceeds the force exerted by the valve spring, the valve disk lifts off its seat, allowing some of the system fluid to escape through the valve outlet until system pressure is reduced to just below the relief setpoint of the valve. The spring then reseats the valve. An operating lever is provided to ally manual cycling of the relief valve or to gag it open for certain tests. Virtually all relief valves are provided with some type of device to allow manual cycling. Other types of relief valves are the high pressure air safety relief valve and the bleed air surge relief valve. Both of these types of valves are designed to open completely at a specified lift pressure, and to remain open until a specific reset pressure is reached, at which time they shut. Many different designs of these valves are used, but the same result is achieved.

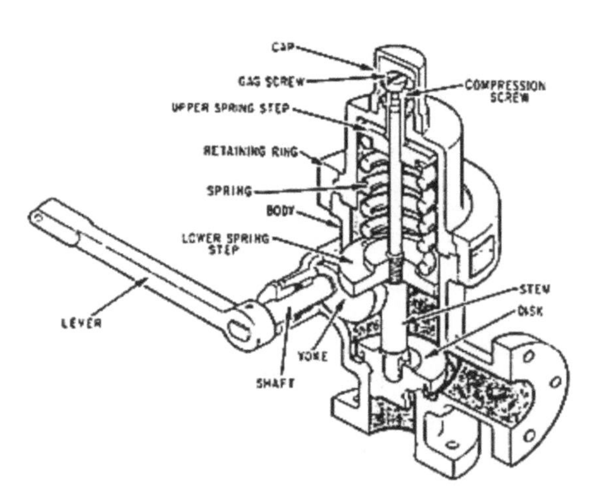

Figure II-58
Spring-loaded Relief Valve

Spring-loaded Reducing Valve

Spring-loaded reducing valves are used in a wide variety of applications. Low-pressure air reducers and others are of this type (see Figure II-59). The valve simply uses spring pressure against a diaphragm to open the valve. On the bottom of the diaphragm, the outlet pressure (the pressure in the reduced pressure system) of the valve forces the disk upward to shut the valve. When the outlet pressure drops below the setpoint of the valve, the spring pressure overcomes the outlet pressure and forces the valve stem downward, opening the valve. As the outlet pressure increases, approaching the desired value, the pressure under the diaphragm begins to overcome spring pressure, forcing the valve stem upwards, shutting the valve. You can adjust the downstream pressure by removing the valve cap and turning the adjusting screw, which varies the spring pressure against the diaphragm. This particular spring-loaded valve will fail in the open position if a diaphragm rupture occurs.

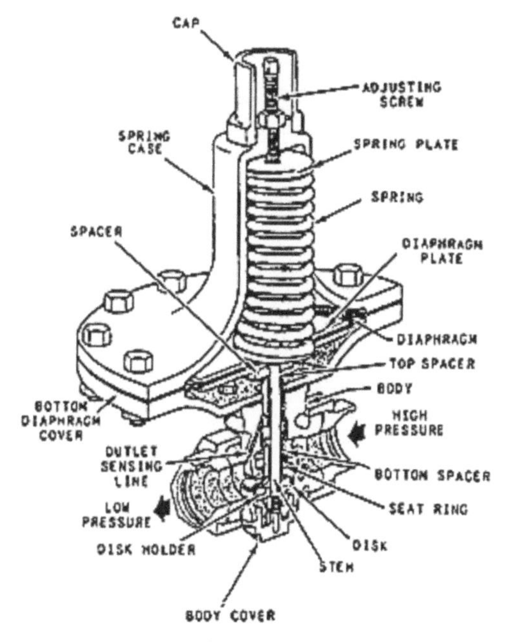

Figure II-59
Pressure Reducing Valve

Section III

Operations and Maintenance Support

CORROSION CONTROL AND TREATMENT

Corrosion can be defined as the deterioration of material by reaction to its environment. Corrosion occurs because of the natural tendency for most metals to return to their natural state; e.g., iron in the presence of moist air will revert to its natural state, iron oxide. Metals can be corroded by the direct reaction of the metal to a chemical; e.g., zinc will react with dilute sulfuric acid, and magnesium will react with alcohols. A properly implemented corrosion control program will disclose corrosion attack in the early stages. Minor maintenance can correct such corrosion.

Basic Causes

Electrochemical corrosion is the most important classification of corrosion. Four conditions must exist before electrochemical corrosion can proceed (see Figure III-1):

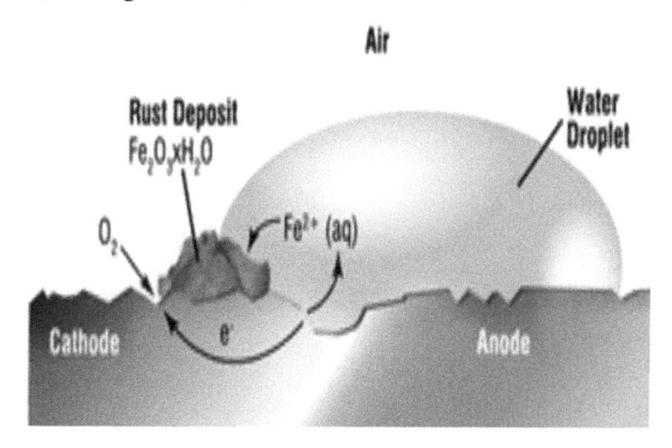

Figure III-1
Water Intrusion Corrosion Process

- There must be something that corrodes (the metal anode).
- There must be a cathode.
- There must be continuous conductive liquid path (electrolyte, usually condensate and salt or other contaminations).
- There must be a conductor to carry the flow of electrons from the anode to the cathode.

This conductor is usually in the form of metal-to-metal contact such as in bolted or riveted joints. The elimination of any one of the four conditions will stop corrosion. An unbroken (perfect) coating on the surface of the metal will prevent the electrolyte from connecting the cathode and anode so the current cannot flow. Therefore, no corrosion will occur as long as the coating is unbroken.

Water Intrusion

Water intrusion is the principal cause of corrosion problems encountered in the field use of equipment. Water can enter an enclosure by free entry, capillary action, or condensation. With these three modes of water entry acting and with the subsequent confinement of water, it is almost certain that any enclosure will be susceptible to water intrusion. Preventive maintenance is the most cost-effective method of controlling corrosion, including problems caused by poor design. See Figure III-2.

Environmental Factors

At normal atmospheric temperatures the moisture in the air is enough to start corrosive action. Oxygen is essential for corrosion to

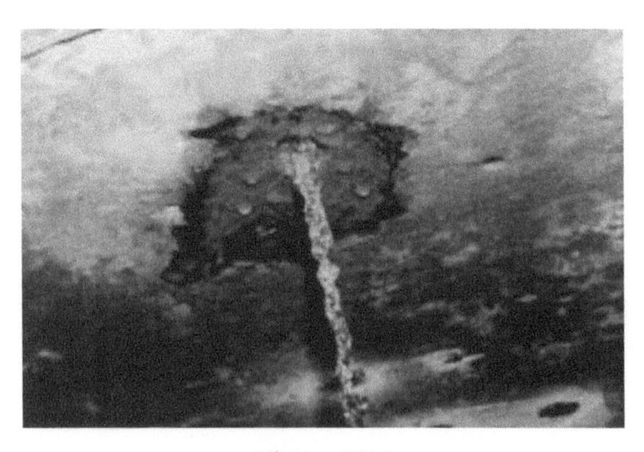

Figure III-2
Worst Case Water Intrusion

occur in water at ambient temperatures. Other factors that affect the tendency of a metal to corrode are:

- acidity or alkalinity of the conductive medium (pH factor)
- stability of the corrosion products
- biological organisms (particularly anaerobic bacteria)
- variation in composition of the corrosive medium
- temperature

Preventive Maintenance

Continuing surveillance is required to disclose corrosion attack in its early stages. Without proper preventive maintenance, corrosion can seriously damage equipment. All equipment must be carefully inspected for signs of corrosion during scheduled and random inspections. These activities should be organized and properly managed to produce an effective program. Materials that require special treatment to protect them against corrosion are those most vulnerable to corrosion attack and require careful inspection and maintenance. Preventive maintenance as related to corrosion control includes the following specific functions:

- An adequate cleaning program.
- Detailed scheduled inspection of facilities and systems for corrosion and failure of protective coating systems.
- Prompt treatment of corrosion after it is detected and touchup of damaged paint areas.
- Periodic lubrication and use of supplementary preservative coatings.
- Adequate drainage of moisture entrapment areas by maintaining drain holes free of obstruction.
- Periodic removal of accumulated water and other foreign matter.
- Coating of exposed critical surfaces with preservative compound.
- Frequent wiping and cleaning of surfaces that must remain bare.
- Use of covers or storage of critical components in protected enclosures.

- Periodic and frequent inspections of areas where absorbent materials are in contact with metals.

Corrosion Removal and Treatment

If the corroded area is soiled by foreign materials such as grease or dirt, the surfaces must be cleaned before stripping paint or removing corrosion. Failure to adequately clean all residues will permit corrosion to continue. Dry abrasive blasting with fine abrasives is also an acceptable method of removing corrosion products when surfaces will subsequently be painted and where dimensional tolerances are not critical. Although abrasive blasting is the preferred method of removing corrosion, other mechanical methods that may be used are grinding, chipping, sanding or wire brushing. All abrasive residues should be thoroughly removed using high-pressure clean air or running water. Never use carbon steel wool or wire brushes because particles from these materials may become imbedded in the surface causing galvanic corrosion problems. Residual paint or primer may be removed by mechanical or chemical treatment after cleaning surfaces. Chemical corrosion removal may be used when there is no danger of the chemical becoming entrapped. The chemical method should be used on complex shapes and machined surfaces.

Five Most Common Methods Used to Control Corrosion

Protective Coatings and Linings

Coatings and linings are principal tools for defending against corrosion. They are often used in conjunction with cathodic protection systems to provide the most cost effective protection for the structure. Coatings and linings help protect against corrosion in three ways:

- They provide a barrier to prevent or limit contact between a structure's metal surface and its corrosive environment;
- They release substances that inhibit the corrosion process and protect the structure from deteriorating;
- They serve as sacrificial materials, such as when galvanizing is used.

To be effective, protective coatings and linings must be properly selected and installed by personnel trained in surface preparation and application of the material selected.

Cathodic Protection

Cathodic protection (CP) is a technology which uses direct electrical current to counteract the normal external corrosion of a structure that contains metal, such as a boat or a ship with steel-reinforcing components. The term "cathodic" refers to the area of the metal where corrosion is controlled, as opposed to the anodic areas where corrosion occurs. The principle behind cathodic protection is to make the entire surface of a structure behave like a cathode with respect to an external anode. This behavior is induced by installing sacrificial materials to serve as anodes or by applying an external direct current power source in conjunction with anodes. On new structures, cathodic protection can help prevent corrosion from starting; on existing structures, cathodic protection can help stop existing corrosion from getting worse. Effective CP system design will take into account such variables as:

- variations in the environment surrounding a structure
- the presence of protective coatings and linings
- the metal to be protected; the expected useful life of the structure
- the ability to maintain the cathodic protection system
- the total electrical current required for protection

The costs of installing and maintaining cathodic protection must be considered in context of the direct expenses associated with replacement of corroded structures and possible structural failure, as well as indirect costs such as environmental damage. Installing cathodic protection on any infrastructure can be very costly.

Materials Selection

Materials selection refers to the selection and use of corrosion-resistant materials such as stainless steels, plastics, and special alloys to enhance the life span of a product. Materials selection personnel consider the environment in which the product will exist and the desired life span. If more than one material is used, such as two metals joined together, controlling corrosion requires that the materials have compatible electrochemical properties. The two most common materials used in consumer products are steel and aluminum, which can be severely affected by corrosion.

Inhibitors, Vapor Corrosion

Vapor-phase corrosion inhibitors (VpCIs) are substances which, when added to a particular environment, decrease the rate of attack of that environment on material such as metal. VpCIs are commonly added in small amounts to liquids such as acids, cooling waters, and steam, either continuously or intermittently, to prevent serious corrosion. Inhibitors can stop or retard corrosion in many ways, such as "adsorption," forming films to coat materials at a molecular level and protect them from attack, and creating a "passive" layer on the surface of a material which inhibits further deterioration. VpCIs can extend the life of equipment, prevent system shutdowns and failures, avoid product contamination, prevent loss of heat transfer, and preserve an attractive appearance of products.

Evaluating the environment in which a structure is or will be located is very important to corrosion prevention, no matter which control method is used. Modifying the environment immediately surrounding a structure, such as reducing moisture or improving drainage, can be a simple and effective way to reduce the potential for corrosion.

Copper and Copper-based Alloys

Protective coatings are normally not required for copper and copper-base alloys because of their inherent corrosion resistance. The green tarnish commonly noted on copper alloys does not normally affect its performance characteristics. The green patina actually provides corrosion protection to the base metal. However, copper-based materials should be protectively coated in highly acidic conditions.

Communications, Electronic and Associated Electrical Equipment

Condensate enters non-airtight components in the form of moist air. If this air is moist and if the temperature of the components drops to the dew point during the temperature cycle, a film of moisture is deposited on the inside of the components. Moisture will therefore accumulate as a result of many temperature cycles. Experience has shown that serious problems, such as corrosion, fungus growth, changes in electrical characteristics, and shorting can occur as a result of the accumulated moisture. Preventive procedures for controlling moisture problems include hermetically sealing, application of a conformal coating, pressurizing with dry gas, ventilation of enclosed areas, use of desiccants, use of volatile corrosion inhibitors, potting of electrical connectors, heating to prevent cycling to the dew point, providing static and dynamic dehumidification systems, and providing adequate drain holes to prevent moisture accumulation. Each problem must be inspected to determine the most practical preventive procedure to follow. However, certain procedures and preventive measures are common to most moisture problems and can be readily performed.

Nuts and Bolts

Failures of nuts and bolts can be reduced by conscientious surface preparation prior to application of a protective coating. A brush coat of primer prior to spray application will ensure adequate coverage. Bolts and nuts should be specified as hot-dipped galvanized where possible.

ELECTRIC ARC FLASH

An arc flash hazard is defined as a dangerous condition associated with the release of energy caused by an electric arc and explosion triggered by an electrical arc (see Figure III-3). Arc flashes occur when conductive objects get too close to exposed current sources. The arc current creates a brilliant flash of light, a loud noise, intense heat, and a rapidly moving pressure wave. The temperature of the electrical arc can cause fatal and major burns at distances of 5-10 feet from the arcing equipment. Dropping tools into an electrical enclosure, opening equipment doors, racking breakers, and even checking voltage can cause an arc flash unless proper safety precautions are taken. Arc flashes can result in:

- exposure to extreme temperatures reaching 5,000°F+ (electrical arc burns make up a substantial portion of injuries from electrical malfunctions)
- blast waves that emit shrapnel, toxic gases, metal vapors, molten metal droplets and intense UV rays (blast products shower the immediate vicinity of the arcing fault)

Definitions of Terms

The following terms taken from the NFPA 70E, 2012 edition, are frequently found on arc flash labels and signs:

Nominal System Voltage (NSV)

The NSV is normally the voltage required by the largest loads in a system. Common industrial values are 120, 208, 220, and 480 volts.

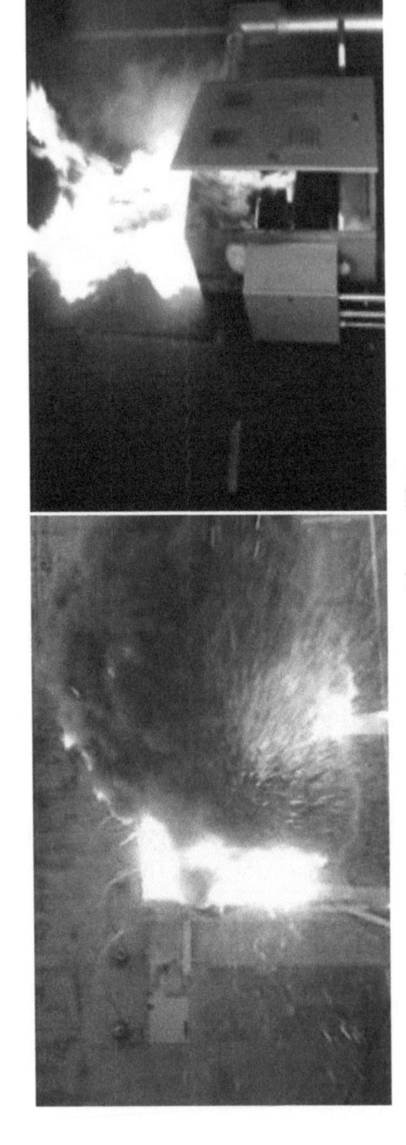

Figure III-3
Electric Arc Flash and Subsequent Fire

This measurement can be VAC or VDC and required by 2012 NFPA 70E (130.5 C) to be displayed on arc flash labeling.

Arc Flash Boundary

The arc flash boundary is the distance at which a person is likely to receive a second-degree burn. The onset of a second-degree burn is possible when the skin receives 1.2 cal/cm^2 of incident energy. Calculations based on 2012 NFPA 70E Annex D.7.5. This measurement is required by 2012 NFPA 70E(130.5 C) to be displayed on arc flash labeling.

Available Incident Energy at a Working Distance

This is the energy per unit area on a surface located at the normal working distance from the potential arc fault. The incident energy is most commonly measured in units of calories per square centimeter. Second-degree burns occur at an energy level of approximately 1.2 cal/cm^2.

Required Level of PPE

The personal protective equipment (PPE) required is dependent on the incident energy at every point a person may perform work on energized equipment. An electrical engineer or other qualified person should cover all parts of the body that may be exposed to an arc flash. This could include boots, gloves, flame resistant clothing, safety glasses, etc. Hearing protection and leather gloves are required for all hazard risk categories (Tables 130.7 C).

Hazard Risk Category (HRC)

A general classification of hazard is involved in performing specified tasks. HRC typically ranges from zero to four, with zero denoting minimum-risk activities and four denoting high-risk activities. The NFPA provides a recommended list of PPE for each HRC in Table 130.7. HRC levels are not associated with a specific measurement of cal/cm^2 by the NFPA, but rather a defined list of PPE.

Shock Protection Boundary

Limited approach boundary (Annex 0.1.1 and 0.1.2.2)—This boundary may only be crossed by a qualified person. An unqualified person, wearing appropriate PPE may cross if accompanied by a qualified person. Becoming qualified requires special training. Restricted approach boundary (Annex C.1.2.3)—This boundary may only be crossed by authorized management (using adequate shock prevention equipment and techniques. Prohibited approach boundary (Annex C.1.2.4) —This boundary may only be crossed by a qualified person that has the same level of protection required for direct contact with live parts. See Figure III-4.

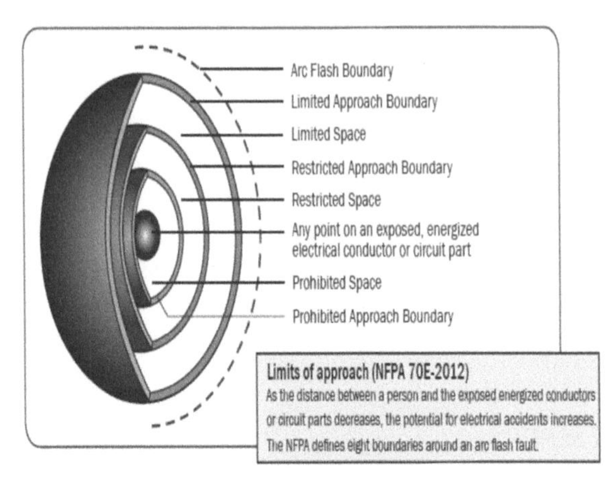

Figure III-4
Safe Distance Boundaries

Communication

A key element in any arc flash safety program is good visual communication. Using labels and signs to warn workers, emergency responders and others of a potential arc flash hazard is critical safety information and saves lives. The NEC (National Electrical Code), Article 110-16/Flash protection states in part that switchboards, panelboards, industrial control panels, and motor control centers that are likely to require examination, adjustment, servicing, or maintenance (while energized) shall be field marked to warn qualified persons of potential electric arc-flash hazards. In addition, proposed changes to NFPA 70E will require that arc flash labels must contain information specific to each piece of equipment, detailing the exact hazards.

> **Note**: Conduct an inspection of your facility to determine the need for new and replacement arc flash labels.

When evaluating your facility, check for proper labeling on all required structures and devices, legibility and accuracy of all existing labels, equipment that isn't carrying labels, new equipment that poses an arc flash danger, boundaries calculations at posted equipment and areas where people should be excluded.

PPE (Personal Protective Equipment)

The personal protective equipment required is dependent on the incident energy at every point a person may perform work on energized equipment. An electrical engineer or other qualified person should perform the calculations that determine the incident energy. Qualified workers should cover all parts of the body that may be exposed to an arc flash. The appropriate PPE includes boots, gloves, flame resistant clothing, safety glasses etc. The potential for injury can be reduced using various electrical safety tools and techniques. Remote breaker racking, remote door opening and closing and wearing the proper personal protective equipment (PPE) all offer improved safety. See Figure III-5.

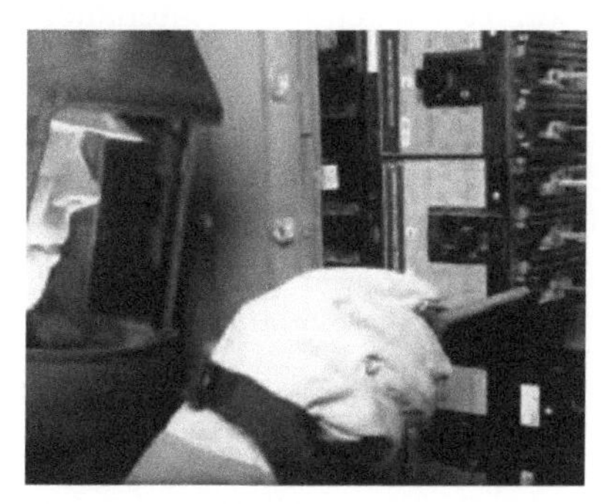

Figure III-5
Frequently Tested and Certified Gloves

FLAMMABLE AND COMBUSTIBLE LIQUIDS

There are two primary hazards associated with flammable and combustible liquids: explosion and fire. Proper storage and use of flammable liquids can significantly reduce the possibility of accidental fires and injuries. To prevent these hazards, two standards address the primary concerns of design and construction, ventilation, ignition sources and storage. To minimize risk to life and property, the requirements of the Uniform Fire Code—(Specifically NFPA 30):

Note: Scope of NFPA 1 and OSHA Standard 1910.106, must be adhered to. Flammable and Combustible Liquids Code 29 CFR 1910.106 applies to the handling, storage, and use of flammable and combustible liquids with a flash point below 200°F.

The scope includes, but is not limited to, the following: (1) Inspection of permanent and temporary buildings, processes, equipment, systems, and other fire and related life safety situations; (2) Investigation of fires, explosions, hazardous materials incidents, and other related emergency incidents; (3) Review of construction plans, drawings and specifications for life safety systems, fire protection systems, access, water supplies, processes, hazardous materials, and other fire and life safety issues; (4) Fire and life safety education of fire brigades, employees, responsible parties, and the general public; (5) Existing occupancies and conditions, the design and construction of new buildings, remodeling of existing buildings, and additions to existing buildings; (6) Design, alteration, modification, construction, maintenance, and testing of fire protection systems and equipment; (7) Access requirements for fire department operations; (8) Hazards from outside fires in vegetation, trash, building debris, and other materials; (9) Regulation and control of special events including, but not limited to, assemblage of people, exhibits, trade shows, amusement parks, haunted houses, outdoor events, and other similar special temporary and permanent occupancies; (10) Interior finish, decorations, furnishings, and other combustibles that contribute to fire spread, fire load, and smoke production; (11) Storage, use, processing, handling, and on-site transportation of flammable and combustible gases, liquids, and solids; (12) Storage, use, processing, handling, and on-site transportation of hazardous materials; (13) Control of emergency operations and scenes; (14) Conditions affecting fire fighter safety; (15) Arrangement, design, construction, and alteration of new and existing means of egress 1.1.2 Title. The title of this Code shall be NFPA 1, Fire Code, of the National Fire Protection Association (NFPA). See Figure III-6.

Definitions

Flammable liquid—any liquid having a flash point below 100°F (37.8°C), except any mixture having components with flashpoints of 100°F (37.8°C) or higher, the total of which make up 99 percent or more of the total volume of the mixture. Flammable liquids shall be known as

Figure III-6
Electrically Grounded Drum and Safety Can

Class I liquids. Class I liquids are divided into three classes as follows:

- Class IA—flashpoint below 73°F and boiling point below 100°F
- Class IB—flashpoint below 73°F and boiling point above 100°F
- Class IC—flash at or above 73°F and below 100°F

It should be mentioned that flash point was selected as the basis for classification of flammable and combustible liquids because it is directly related to a liquid's ability to generate vapor; i.e., its volatility. Because it is the vapor of the liquid, not the liquid itself that burns, vapor generation becomes the primary factor in determining the fire hazard. The expression "low flash/high hazard" applies. Liquids having flash points below ambient storage temperatures generally display a rapid rate of flame spread over the surface of the liquid, because it is not necessary for the heat of the fire to expend its energy in heating the liquid to generate more vapor.

Combustible liquid—any liquid having a flash point at or above 100°F (37.8°C). Combustible liquids shall be divided into two classes as follows:

- Class II Combustibles—Flashpoint above 100°F and below 140°F
- Class III Combustibles—Flashpoint at or above 140°F
  - Subclass IIIA—flashpoint at or above 140°F and below 200°F
  - Subclass IIIB—flashpoint at or above 200°F

Note: When a combustible liquid is heated to within 30°F (16.7°C) of its flash point, it shall be handled in accordance with the requirements for the next lower class of liquids.

Containers

The capacity of flammable and combustible liquid containers will be in accordance with Table III-1. There is flexibility in these requirements now, as fire suppression systems and types of construction can multiply these capacities by two or three fold.

Table III-1
Flammable and Combustible Liquid Container Capacity

	Flammable Liquids		Combustible Liquids		
Container	**1A**	**1B**	**1C**	**II**	**III**
Glass or approved plastic1	1 pt2	1 qt2	1 gal	1 gal	1 gal
Metal (Other than DOT	1 gal	5 gal	5 gal	5 gal	5 gal
Safety Cans	2 gal	5 gal	5 gal	5 gal	5 gal
Metal drums (DOT	60 gal	60 gal	60 gal	60 gal	60 gal
Approved portable tanks	660 gal	660 gal	660 gal	660 gal	660 gal

Design, Construction and Capacity of Storage Cabinets

Not more than 60 gallons of Class I and/or Class II liquids, or not more than 120 gallons of Class III liquids may be stored in an individual cabinet. This standard permits both metal and wooden storage cabinets. Storage cabinets shall be designed and constructed to limit the internal temperature to not more than 325°F when subjected to a standardized 10-minute fire test. All joints and seams shall remain tight and the door shall remain securely closed during the fire test. Storage cabinets shall be conspicuously labeled, "Flammable—Keep Fire Away." The bottom, top, door and sides of metal cabinets shall be at least No. 18 gage sheet metal and double walled with 1½-inch air space. The door shall be provided with a three-point lock, and the door sill shall be raised at least 2 inches above the bottom of the cabinet. See Figure III-7.

Figure III-7
Properly Labeled Storage Cabinet

Storage in inside storage rooms shall comply with the following, as shown in Table III-2:

In every inside storage room, there shall be maintained an aisle at least 3 feet wide. Easy movement within the room is necessary, to reduce the potential for spilling or damaging the containers and to provide both access for firefighting and a ready escape path for occupants of the room, should a fire occur. Containers over 30 gallons capacity shall not be stacked one upon the other. Such containers are built to DOT specifications and are not required to withstand a drop test greater than 3 feet when full. Dispensing shall be only by approved pump or

Table III-2
Storage in Inside Rooms

Fire Protection Provided[1]	Fire Resistance	Maximum Floor Area (ft²)	Total Allowable Quantities (gal/ft² floor area)
Yes	2 hr.	500	10
No	2 hr.	500	4*
Yes	1 hr.	150	5*
No	1 hr.	150	2*

self-closing faucet.

Note: The quantity of liquid that may be located outside of an inside storage room or storage cabinet in a building or in any one fire area of a building shall not exceed:

- 25 gallons of Class IA liquids in containers
- 120 gallons of Class IB, IC, II, or III liquids in containers
- 660 gallons of Class 1B, 1C, II, or III liquids in a single portable tank.

Ventilation

Every inside storage room will be provided with a continuous mechanical exhaust ventilation system. To prevent the accumulation of vapors, the location of both the makeup and exhaust air openings will be arranged to provide, as far as practical, air movement directly to the exterior of the building, and if ducts are used, they will not be used for any other purpose.

GREASE AND OIL LUBRICATION

Tribology

Tribology is the study and application of the principles of friction, lubrication and wear. As a branch of mechanical engineering, it deals with the analysis and application of the physics associated with friction, the loss of it (lubrication) and its destructive properties (wear). This "refresher" is primarily focused on how proper grease and oil lubrication increases equipment uptime, reduces mechanical wear and equipment failures, while enabling more effective application of maintenance budgets and ensuring prolonged equipment life cycles. It explores types of lubricating conditions, forms and types of lubricants and additives and their comparative properties.

Friction

Friction is an essential yet often overlooked concept in physics and a key factor in how mechanisms work. It is also the most relevant property in oil analysis and an important aspect for measuring oil's consistency, efficiency, viscosity and the contaminants in it. Mechanical wear, caused by surface degradation, including fatigue, costs facilities billions of dollars each year to repair.

Lubricants

Though the ability to minimize friction is the number one function of a lubricant, there are other major functions that must be considered. A lubricant is likely to also be required to:

- *Clean*—A lubricant must maintain internal cleanliness by suspending contaminants or keeping contaminants from adhering to components.
- *Cool Moving Elements*—Reducing friction will reduce the amount of heat that is generated and lower the operating temperature of the components. A lubricant must also absorb heat from the components and transport it to a location where it can be safely dissipated.
- *Prevent Contamination*—The lubricant should act as a dynamic seal in locations such as the piston, piston ring and cylinder contact areas. This minimizes contamination by combustion byproducts (for example) in the lubricating system. Lubricants are also relied upon to support mechanical seals found elsewhere and to minimize external contamination and fluid loss.
- *Dampen Shock*—The lubricant may be required to cushion the blows of mechanical shock. A lubricant film can absorb and disperse energy spikes over a broader contact area.
- *Transfer Energy*—A lubricant may be required to act as an energy transfer median as in the case of hydraulic equipment or lifters in an automotive engine.
- *Prevent Corrosion*—A lubricant must also have the ability to prevent or minimize internal component corrosion. This can be accomplished either by chemically neutralizing the corrosive products or by setting up a barrier between the components and the corrosive material.

Grease

According to the *Practical Handbook of Lubrication*, grease is a lubricant composed of a fluid lubricant thickened with a material that contributes a degree of plasticity. Greases are typically used in areas where a continuous supply of oil cannot be retained, such as open bearings or chassis components.

Components

Greases are comprised of three essential components: a base fluid, a thickening system and an additive system which are combined to give the final product its special lubricating properties. Different types of base oils may be used in the manufacture of grease, including petroleum (naphthenic, paraffinic) and synthetic (PAOs, esters, silicones, glycols). Viscosity of the base oil is its most significant property. Lighter, lower viscosity base oil is used to formulate low temperature greases or greases suitable for high rotational speeds, while heavier, higher viscosity base oils are used to formulate greases used in applications where high loading is encountered, high temperature, and or low rotational speeds are seen.

Thickeners

Thickeners are ingredients added to base oil to thicken it to a grease structure. The two basic types of thickeners are organic and inorganic. To take on enhanced performance characteristics, including higher dropping points, a complex agent is added to the soap thickener to convert it to a soap salt complex thickener. The greases are then referred to as "complexes" and include the most popular thickener system, lithium complex, as well as aluminum complex and others.

Additives

Chemical and metallic additives are added to greases to enhance their performance, much like the additives added to lubricating oils. Performance requirements, compatibility, environmental considerations, color and cost are all factors of additive selection. Greases may be categorized according to their thickener system, for example lithium complex or bentone, or polyurea; their metallic or solid additive constituent, for example moly or Teflon; their performance characteristics, such as high temp, low temp, impact resistant, or by application, i.e. water pump, wheel bearing, chassis, etc.

Properties

Grease consistency correlates to the firmness of the grease. Depending on the applications they're designed for, greases can range from semi-fluid consistencies to almost solid. Care must be taken to select the correct consistency for the application. If the grease is too hard, it may not adequately flow to the areas in need of lubrication. If it is too soft, it may leak away from the desired area. Since consistency directly correlates to pump-ability, equipment greased through a dispensing system may require a grease representing a compromise between what is required for lubrication and what can be adequately pumped by the hardware used.

> **Note**: Over-greasing can lead to high operating temperatures, collapsed seals and, in the case of greased electric motors, energy loss and failures. The best ways to avoid these problems are to establish a maintenance program, use calculations to determine the correct lubricant amount and frequency of re-lubrication, and utilize feedback instruments.

Tips to Control Over-greasing

- Discontinue greasing when you feel abnormal back pressure.
- Always make sure exhaust ports are cleaned of any debris or old, hard crust that could be blocking the passageway.
- Consider installing grease guns with pressure gauges, shut-off grease fittings or relief-type vent plugs.
- Slowly pump grease into bearings every few seconds. Using a quick-lever action could cause seal damage and not allow the grease to distribute throughout the bearing correctly.

Oil

Lubricating oils encompass a large group of emollients that serve numerous industrial and residential purposes. They are classified into various types, depending on their viscosity grades and use of additives. Each type of oil has unique properties, specification requirements, quality requirements and is intended for a special use.

Characteristics

Viscosity (weight) is the oil's resistance to flow or motion. Viscosity varies under different temperatures and is vital for maintaining a lubricant film between moving parts. Viscosity plays a role in an engine's cold-cranking ability, the movement of gears, meeting load capacities, heat-up of critical engine parts, and the oil consumption rate. Straight weight oil viscosity levels (5, 10, 15, 20, 30, 40, 50, etc.) have a number, often followed by the letter "W." The "W" signifies the low-temperature pour character of the oil and also its minimum viscosity at 212°F.

Properties

- *Anti-Foam Protection*: Foam can lead to oxidation and the inability of oil to create a protective film.
- *Anti-Wear Agents*: These additives decrease wear of heavily stressed components like valve train pieces.
- *Corrosion Protection*: Short operating intervals and slow warm-ups create acid in engine oil. These by-products of the combustion process can damage metal engine parts. Chemical additives are used to neutralize these acids.
- *Film Strength*: The strength of an oil (its ability to keep metal pieces apart) is critical to engine survival.
- *Flash Points*: The lowest temperature at which vaporized oil will flash when exposed to high heat or flame.
- *Lubricity*: The slipperiness or lubricity of oils determines their ability to reach parts and stay in place to prevent wear.
- *Pour Points*: An oil's ability to pour at ultra-low temperatures solely by gravity.
- *Seal Swell Agent*: Swells electrometric seals in a controlled manner to prevent seal leaks.
- *Shear*: Oil lubricating quality. The ability of oil to remain stable, intact, and functional under load.
- *Thermal Stability*: Ability to handle high temperatures without breaking down or forming carbonaceous deposits.

MACHINE GUARDING CHECKLIST

Note: Answers to the following questions should help determine the safeguarding needs of your workplace, by drawing attention to hazardous conditions or practices requiring correction.

Requirements for All Safeguards

- Do the safeguards provided meet the minimum OSHA requirements?
- Do the safeguards prevent workers' hands, arms, and other body parts from making contact with dangerous moving parts?
- Are the safeguards firmly secured and not easily removable?
- Do the safeguards ensure that no object will fall into the moving parts?
- Do the safeguards permit safe, comfortable, and relatively easy operation of the machine?
- Can the machine be oiled without removing the safeguard?
- Is there a system for shutting down the machinery before safeguards are removed?
- Can the existing safeguards be improved?

Mechanical Hazards

Point of Operation

- Is there a point-of-operation safeguard provided for the machine?
- Does it keep the operator's hands, fingers, body out of the danger area?
- Is there evidence that the safeguards have been tampered with or removed?
- Could you suggest a more practical, effective safeguard?
- Could changes be made on the machine to eliminate the point-of-operation hazard entirely?

Power Transmission Apparatus

- Are there any unguarded gears, sprockets, pulleys, or flywheels on the apparatus?
- Are there any exposed belts or chain drives?
- Are there any exposed set screws, key ways, collars, etc.?
- Are starting and stopping controls within easy reach of the operator?

- If there is more than one operator, are separate controls provided?

Other Moving Parts

- Are safeguards provided for all hazardous moving parts of the machine including auxiliary parts?

Non-mechanical Hazards

- Have appropriate measures been taken to safeguard workers against noise hazards?
- Have special guards, enclosures, or personal protective equipment been provided, where necessary, to protect workers from exposure to harmful substances used in machine operation?

Electric Hazards

- Is the machine installed in accordance with National Fire Protection Association and National Electrical requirements?
- Are there loose conduit fittings?
- Is the machine properly grounded?
- Is the power supply correctly fused and protected?
- Do workers occasionally receive minor shocks while operating any of the machines?

Training

- Do operators and maintenance workers have the necessary training in how to use the safeguards and why?
- Have operators and maintenance workers been trained in where the safeguards are located, how they provide protection, and what hazards they protect against?
- Have operators and maintenance workers been trained in how and under what circumstances guards can be removed?
- Have workers been trained in the procedures to follow if they notice guards that are damaged, missing, or inadequate?

Protective Equipment and Proper Clothing

- Is protective equipment required?

- If protective equipment is required, is it appropriate for the job, in good condition, kept clean and sanitary, and stored carefully when not in use?
- Is the operator dressed safely for the job (i.e., no loose-fitting clothing or jewelry)?

Machinery Maintenance and Repair

- Have maintenance workers received up-to-date instruction on the machines they service?
- Do maintenance workers lock out the machine from its power sources before beginning repairs?
- Where several maintenance persons work on the same machine, are multiple lockout devices used?
- Do maintenance persons use appropriate and safe equipment in their repair work?
- Is the maintenance equipment itself properly guarded?
- Are maintenance and servicing workers trained in the requirements of 29 CFR 1910.147, lockout/tag-out hazard, and do the procedures for lockout/tag-out exist before they attempt their tasks?

Source: Occupational Safety and Health Administration OSHA 3067

MAINTENANCE PLANNING AND SCHEDULING

Maintenance planning and scheduling (of work orders) is the hub of a well-conceived and fully functioning maintenance organization. But, the truth be known, work planners themselves don't spend the majority of their time simply planning work. As a general rule, only 10% of their time is actually spent performing true planning. The bulk of their remaining working hours can be broken down as follows:

- 10% determining what parts, materials and devices are needed
- 10% physically locating pricing and ordering those items
- 20% checking reorder quantities for stores; receipt and distribution

- 20% ordering repairs for equipment from the engineering division
- 30% performing a variety of other tasks associated with their positions

A properly planned job will have the needed equipment, parts, materials and tools identified, and availability confirmed, prior to being scheduled. Once the jobs are planned, the craft and labor hour requirements to complete the jobs will be known, so that the correct number of qualified personnel can be assigned, at the proper time. This process is often compromised by unscheduled breakdowns in function and communication. (Unplanned jobs or jobs resulting from equipment breakdown contribute to a maintenance worker's day). Interruptions to smooth work flow can also be caused by lack of adequate planning and scheduling techniques. Typical maintenance work interruptions include:

- Lack of adequate supplies
- Trips to and from the storeroom for parts
- Lack of information regarding job requirements
- No shutdown coordination
- Trips to and from the shop for special tools
- Waiting for ancillary equipment availability
- Waiting for another craft

Planned Maintenance

Planned maintenance releases work leaders from major planning duties and allows them more time to supervise their crews. It provides procedures to plan, execute, monitor and control maintenance resources; reduces delays in waiting for manpower, material and tools after a job is in progress; provides for systematic collection of materials prior to planned jobs; provides procedures to implement and continue a PM program; provides a communication link between maintenance and operations; provides a daily plan for work leaders; allows hourly employees to be 100% work loaded; helps field repairs coordinate work with shop and construction forces; enables performance reporting, allowing upper management to judge maintenance progress; reduces the time required for critical shutdowns or overhauls; reduces maintenance costs; provides a tool for operations to assign priorities and reduces the incidents of emergency breakdowns. Planning and scheduling functions are the key deliverables of the planning role. This is where the most gains in execution have the potential to be made and acted upon.

Basic Approach

The best programs include well written procedures with data entry forms and management spot checks. Early detection by supervision and operators is key to implementing proper maintenance. Equipment problems, such as leaks, temperature anomalies, and odd sounds should be noted and reported to the work planner to assist with the planning and scheduling process. No problem should be considered too small to report. There are three basic steps to achieving a planned approach to maintenance: the identification and documentation of the work that needs to be performed, the development of a structured process for planning and scheduling that work, and the modification and expansion of those plans based on lessons learned and new requirements. If defined early enough, work can be properly planned. However the planning process needs to be monitored and controlled. The effectiveness of a planning effort is based on economics. Equipment downtime and lost labor time can be reduced by ensuring that materials, tools and equipment are made available before the job starts.

Work Scheduling

It goes without saying that only planned work can be scheduled. Required resources (people, equipment, tools and parts) must be determined and an estimate of the job's duration must be known before the work can be scheduled. A schedule without job plans is just a wish list. When PM (preventive maintenance), PDM (predictive maintenance) and planned work make up the large majority of the workday, the planning is already completed, allowing the work planner to concentrate on developing work plans for the remaining corrective type work orders. Breakdowns and unscheduled downtime are dramatically reduced; stores' inventory can be minimized because supplies and materials required for PM and planned jobs are defined ahead of time; work can be completed more efficiently without interruption; the working climate is improved and the morale of the work force is boosted; and the budgeting process is more streamlined.

Work Orders

Every work order (sometimes called a job order or service order) should be planned before execution. This includes all preventive maintenance work, as well as repairs and route checks. It is very important that adequate personnel be dedicated to planning work full time. If this is not done, planners will often get pulled into the role of supervisor, craftsperson or parts chaser. None of these activities will increase the efficiency of maintenance activities by the magnitude that planning will accomplish. The role of the planner needs to cover the full range of the work order system, from input into coding, prioritization and a degree of autonomy in execution. Each planned job is accompanied by a work package, which is a written document containing all information needed to execute the work. The work package should include:

- A clear scope of the work required
- An accurate estimate of the manpower required
- A detailed procedure for performing the work
- A complete list of all tools and equipment required
- All non-standard tools acquired and staged at work site
- A detailed parts list
- The location of the parts, staged near the work area in kit form
- All necessary permits attached
- Sketches, drawings, digital photographs as necessary
- Contact information, should questions arise
- Special notes, instructions
- Coordinated vendor support, etc.
- Schedule for execution for each craft, production, etc.
- Safety and environmental hazard communication
- Personal protective equipment (PPE) required

Labor Estimates

Effective planning involves how best to estimate the labor hours for a job plan. Estimated labor hours enable scheduling and promote productivity. Due to the nature of maintenance work and the individuals' knowledge of the work to be performed, work planners often rely on work histories and/or pre-established engineering standards. A skilled technician can usually make just as accurate an estimate (plus or minus 100 percent) from a simple review of the job and a quick review of the equipment file (if any history exists). The planner with a skilled technician background should estimate labor hours for a smooth job by a qualified technician. The simple estimate of a planner with significant craft experience is preferred over other methods. It yields an estimate useful enough to provide support for scheduling.

Shutdowns

Shutdowns (both scheduled and unscheduled) can seriously impact overall maintenance operations. There are three basic concepts that govern how to manage the planning and scheduling of a shutdown. These concepts are: how well the shutdown work is planned, what type of work is executed during the shutdown and when the shutdown work list is finalized. Effective shutdown management is critical to the operation of most facilities. Without well-planned and executed shutdowns, equipment reliability suffers, and the plant pays the price with poor quality and lost production. Becoming proficient at managing shutdowns is the path to reducing overall downtime costs, so the shutdown itself does not consume the savings it is capable of generating. It is an exercise in waste reduction. The actions needed to make shutdowns more cost effective can be taken immediately if the principles of successful shutdown management are clearly understood. Effective maintenance planning and scheduling improves maintenance efficiency and increases productivity!

MECHANICAL ROOM SAFETY INSPECTION

A building's mechanical spaces house its heating, ventilation and air conditioning (HVAC) and facility support systems. This can include central utility plants, boiler and chiller rooms, mechanical and electrical rooms and fuel rooms. A mechanical equipment room is a room that houses a variety of mechanical and/or electrical equipment, especially that used to control the environment in a building. The specific equipment found in a mechanical room depends on what is needed for the particular building in which it is located, such as air handling units, heat transfer coils, pumps, fans, motors, and numerous other items that may also be found there. Mechanical room design considerations include provisions for appropriate equipment layout, effective drainage and sufficient ventilation. Safety issues and upkeep of a mechanical room are of prime importance as is security against unauthorized tampering with the equipment inside. Every facility should have a comprehensive program in place to ensure that these aspects are met. Facility managers should conduct regular inspections of their mechanical spaces, and all personnel should be cognizant of their conditions. Some items to look for during a walkthrough might include:

Inspection Checklist

Room Requirements

- Ensure that exhaust outlets are located no closer than 20 feet from inlet vents, and the room is adequately ventilated.
- Restrict access to the mechanical room, and keep the door closed and tightly sealed to isolate the room.
- Maintain the cleanliness of all pits (low areas) in the room to avoid accumulations of chemical residues such as acids, spilled refrigerant, cleaning solvents, etc.
- Maintain the floor drains for free flow and proper point of discharge.
- Keep all work surfaces, walls and floors tidy, dry and clear.
- Utilize safety treads or abrasive mastics on all walking surfaces.
- Ensure proper lighting levels, especially over workstations, stairs and in corridors.
- Ensure any see-through (vision panel) is free from obstruction.
- Ensure the structural integrity of floors, grates, door hardware and handrails.

Safety Related

- Locate self-contained breathing apparatus (SCBA) outside of the mechanical room. In an emergency, a properly used SCBA can mean the difference between life and death, but not if it's stored inside the room where the contamination is likely to occur.
- Identify all asbestos-containing materials.
- Ensure that a first aid box is fully equipped and accessible to staff and that all accidents, near misses and illnesses caused by work are reported and recorded.
- Ensure that all signage is proper, maintained and unobstructed.
- Ensure that all eye wash stations and emergency showers are inspected and functional.
- Ensure that hearing protection, protective clothing and other PPE is available as appropriate.

Equipment Related

- Make certain that all machine guards are in place and stable.
- Restrict access to live high voltage equipment to authorized people only.
- Regularly service all gas appliances and educate personnel regarding gas leak response.
- Maintain equipment in a clean and dry condition.
- Utilize formalized procedures for taking equipment out of service.
- Ensure that there are adequate clearances around equipment and panels.

Piping and fittings

- Ensure piping is labeled with content and flow direction.
- Ensure that discharge lines are the right size, length and properly directed.
- Do not allow fluids to discharge onto the floor or into the flow of traffic.
- All purge vents should be located outside.

Materials and Substances

- Clearly label all chemical containers with contents, hazards warnings and the precautions to be taken (ensure that MSDS sheets are posted nearby for all applicable substances resident in the area and that training is provided in safe use of chemicals and on what to do in an emergency—spillage, poisoning, splashing, etc.).
- Gas cylinders should be stored upright and securely strapped in place.

Fire Code Considerations

Housekeeping

- Properly store combustible materials in mechanical/electrical rooms.
- Do not store combustible materials in exits or stairwells.
- Do not store combustible materials within 2 feet of the ceiling or 18 inches below sprinkler heads.
- Do not store anything within 3 feet of electrical panels, heat-producing equipment or fire-protection equipment.

Exits

- Maintain exits and exit corridors clear and unobstructed.

- Keep mechanical room doors locked at all times.
- Properly illuminate all exit signs..
- Maintain the operability of emergency lights.

Fire Protection Equipment
- Service fire alarm systems.
- Maintain fire sprinkler systems.
- Label fire protection equipment rooms.
- Mount the proper type of fire extinguishers.
- Inspect and service fire extinguishers.

General
- Do not use extensions cords for permanent power or operations.
- Do not daisy chain power strips and extensions cords.
- Fill and label empty slots in electrical panels.
- Properly store flammable and combustible liquids.

NON-DESTRUCTIVE TESTING

I was debating whether to put this subject in the operations section or the engineering section that follows. Either section would merit its entry, since the charts are both engineering and operations relevant. I decided to put it here, to balance the number of items I have included in each of them (14). Non-destructive testing (NDT) techniques are comprised of test methods used to examine, locate and characterize discontinuities within an object, material or system without impairing its functionality. Because it allows inspection without interfering with a device's final use, NDT provides an excellent balance between quality control and cost-effectiveness. Acoustic emission testing (ultrasound), used to monitor changes in a pressure vessel's integrity during hydrostatic testing, is an example of non-destructive testing. Use of these techniques increases operational reliability in the plant—preventing accidents. See Figure III-8.

- Ensuring customer safety
- Enabling process control
- Improving safety awareness
- Improving quality control
- Decreasing equipment failures

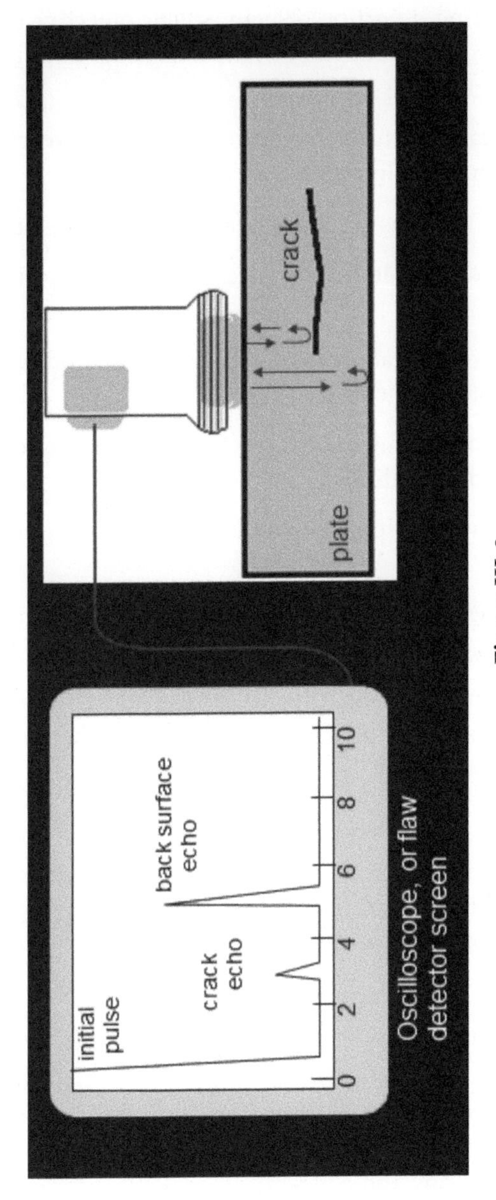

Figure III-8
Ultrasonic Inspection (Pulse-echo)

Categories

Following is a list of non-destructive methods and techniques and their corresponding detection objectives. The objective of each test method is to provide information about the following material parameters:

- Structure or mal-structure (including crystalline structure, grain size, segregation, misalignment).
- Discontinuities (such as cracks, voids, inclusions, de-laminations).
- Dimensions and metrology (thickness, diameter, gap size, discontinuity size).
- Physical and mechanical properties (reflectivity, conductivity, elastic modulus, sonic velocity).
- Composition and chemical analysis (alloy identification, impurities, elemental distributions).
- Stress and dynamic response (residual stress, crack growth, wear, vibration).
- Signature analysis (image content, frequency spectrum, field configuration).
- Color, cracks, dimensions, film thickness, gauging, reflectivity, strain distribution and magnitude, surface finish, surface flaws, through-cracks.

Penetrating Radiation

Covers cracks, density and chemistry variations, elemental distribution, foreign objects, inclusions, micro-porosity, misalignment, missing parts, segregation, service degradation, shrinkage, thickness and voids.

Electromagnetic and Electronic

Include alloy content, anisotropy, cavities, cold work, local strain, hardness, composition, contamination, corrosion, cracks, crack depth, crystal structure, electrical and thermal conductivities, flakes, heat treatment, hot tears, inclusions, ion concentrations, laps, lattice strain, layer thickness, moisture content, polarization, seams, segregation, shrinkage, state of cure, tensile strength and thickness. See Figure III-9.

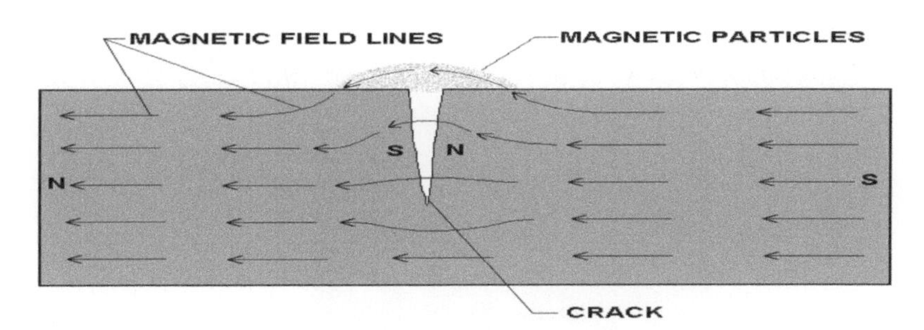

Figure III-9
Surface Crack Detected Using the Magnetic Flux Method

Sonic and Ultrasonic

Includes crack initiation and propagation, cracks, voids, damping factor, degree of cure, degree of impregnation, degree of sintering, de-laminations, density, dimensions, elastic moduli, grain size, inclusions and mechanical degradation.

Thermal and Infrared

Covers, bonding, composition, emissivity, heat contours, plating thickness, porosity, reflectivity, stress, thermal conductivity, thickness and voids.

Chemical and Analytical

Encompass alloy identification, composition, cracks, elemental analysis and distribution, grain size, inclusions, macrostructure, porosity, segregation and surface anomalies. See Figure III-10.

Image Generation

Involves, dimensional variations, dynamic performance, anomaly characterization and definition, anomaly distribution, anomaly propagation and magnetic field configurations.

Signal Image Analysis

Involves data selection, processing and display, anomaly, mapping, correlation and identification, image enhancement, separation of multiple variables and signature analysis.

Advantages and Limitations

Properly utilized, these non-invasive tools can be valuable additions to your maintenance arsenal, used for diagnosing a plethora of

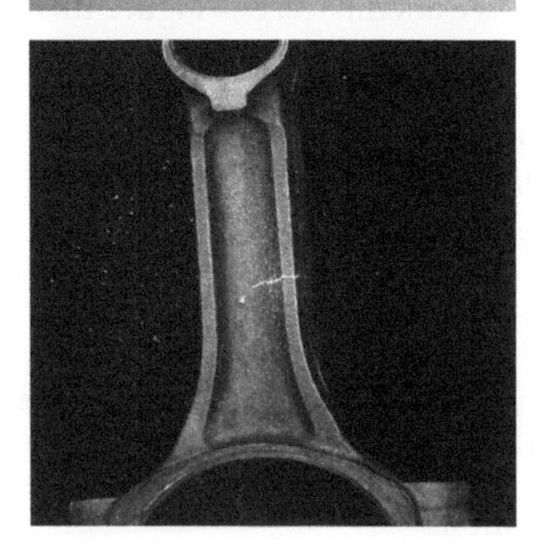

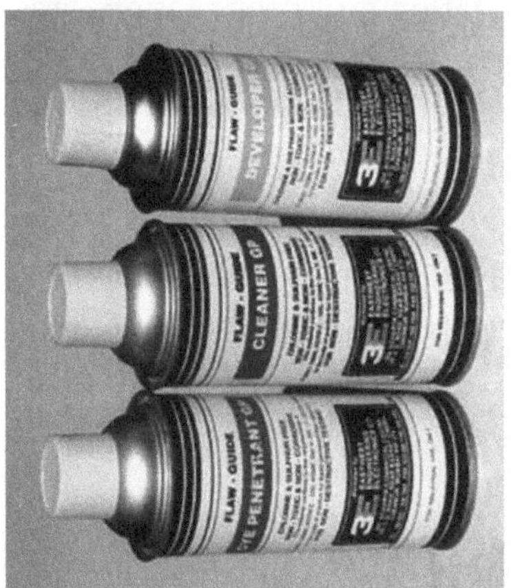

Figure III-10
Fluorescent Penetrant Indications (Dye-penetrant Test)

problems and conditions—checking for steam line, air and water leaks, plate thicknesses of heat exchangers, pre-failure surface cracks in pressure vessels and tube wall ligaments, cracks in pump impellers and myriad other anomalies. See Table III-3.

Costs, Requirements and Capabilities

Maintaining the systems and equipment within a physical plant calls for more than just a cursory mechanical ability. Mechanics should also have a working understanding of physical laws (physics), proven troubleshooting skills and a familiarity with the operational aspects of the infrastructure and devices in their charge. With that arsenal, they can better benefit from the utilization of NDT tools and techniques. Table III-4 is a list of items to consider when determining which, if any of the listed test methods, they may choose to use.

PACKING PUMPS AND VALVES

Installing and adjusting pump and valve packing is both an art and a science. The importance of packing glands correctly cannot be over-emphasized. Many packing failures are due to incorrect installation of the packing. Effectively sealing a pump or valve with packing material is dependent upon those tools and techniques, as well as the overall condition of the device components. Specific tools are required for removal of the old packing and installation of the new packing, as well as tensioning of the fasteners. Always use standard safety equipment and follow good safety practices. The following equipment should be acquired prior to installation:

Calibrated Packing Ring Cutter

- Calibrated torque wrench
- Flashlight
- Helmet
- Inside & outside calipers
- Lubricant for fasteners
- Mirror
- Packing extractor
- Packing knife
- Safety goggles

Table III-3
Non-destructive Testing Methods and Uses

Method	Characteristics detected	Advantages	Limitations	Example of use
Ultrasonics	Changes in acoustic impedance caused by cracks, nonbonds, inclusions, or interfaces	Can penetrate thick materials; excellent for crack detection; can be automated	Normally requires coupling to material either by contact to surface or immersion in a fluid such as water. Surface needs to be smooth.	Adhesive assemblies for bond integrity; laminations; hydrogen cracking
Radiography	Changes in density from voids, inclusions, material variations; placement of internal parts	Can be used to inspect wide range of materials and thicknesses; versatile; film provides record of inspection	Radiation safety requires precautions; expensive; detection of cracks can be difficult unless perpendicular to x-ray film.	Pipeline welds for penetration, inclusions, and voids; internal defects in castings
Visual optical	Surface characteristics such as finish, scratches, cracks, or color; strain in transparent materials; corrosion	Often convenient; can be automated	Can be applied only to surfaces, through surface openings, or to transparent material	Paper, wood, or metal for surface finish and uniformity
Eddy current	Changes in electrical conductivity caused by material variations, cracks, voids, or inclusions	Readily automated; moderate cost	Limited to electrically conducting materials; limited penetration depth	Heat exchanger tubes for wall thinning and cracks
Liquid penetrant	Surface openings due to cracks, porosity, seams, or folds	Inexpensive, easy to use, readily portable, sensitive to small surface flaws	Flaw must be open to surface. Not useful on porous materials or rough surfaces	Turbine blades for surface cracks or porosity; grinding cracks
Magnetic particles	Leakage magnetic flux caused by surface or near-surface cracks, voids, inclusions, or material or geometry changes	Inexpensive or moderate cost, sensitive both to surface and near- surface flaws	Limited to ferromagnetic material; surface preparation and post-inspection demagnetization may be required	Railroad wheels for cracks; large castings

Table III-4
Test Method Considerations

Important considerations	Test method				
	Ultrasonics	X-ray	Eddy current	Magnetic particle	Liquid penetrant
Capital cost	Medium to high	High	Low to medium	Medium	Low
Consumable cost	Very low	High	Low	Medium	Medium
Time of results	Immediate	Delayed	Immediate	Short delay	Short delay
Effect of geometry	Important	Important	Important	Not too important	Not too important
Access problems	Important	Important	Important	Important	Important
Type of defect	Internal	Most	External	External	Surface breaking
Relative sensitivity	High	Medium	High	Low	Low
Formal record	Expensive	Standard	Expensive	Unusual	Unusual
Operator skill	High	High	Medium	Low	Low
Operator training	Important	Important	Important	Important	
Training needs	High	High	Medium	Low	Low
Portability of equipment	High	Low	High to medium	High to medium	High
Dependent on material composition	Very	Quite	Very	Magnetic only	Little
Ability to automate	Good	Fair	Good	Fair	Fair
Capabilities	Thickness gaging: some composition testing	Thickness gaging	Thickness gaging; grade sorting	Defects only	Defects only

- Steel rule
- Tamping tool
- Vernier dial gauge

Note: Before proceeding, make sure that the unit to be repacked has been properly isolated according to site or plant rules. Most of these guidelines are common to both pumps and valves; however, equipment-specific guidelines are noted as required.

Clean and Examine

- Loosen the gland follower nuts slowly and lift the follower to release any trapped pressure under the packing set.
- Remove all of the old packing from the stuffing box; clean the box and shaft thoroughly and examine the shaft and sleeve for wear and scoring. (Replace the shaft or sleeve if wear is excessive.)

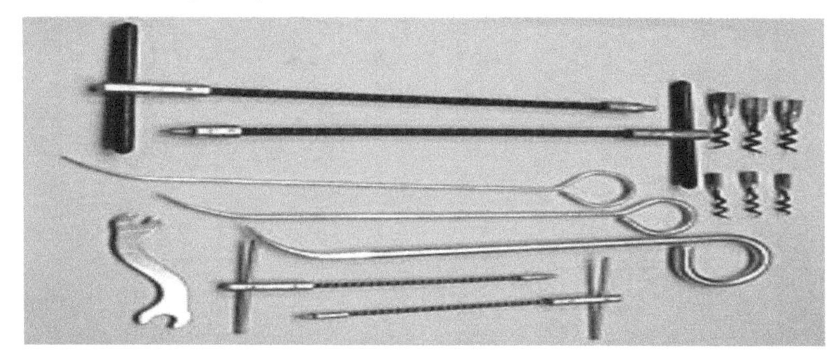

Figure III-11
Packing Extraction Tools

- Examine the shaft and stem for burrs, cracks, corrosion, nicks, scoring or wear that could reduce the packing life.
- Check the stuffing box for excessive clearances and the shaft/stem for eccentricity.
- Replace any defective components and, if in doubt, seek advice.
- Inspect the old packing as part of a failure analysis as to the cause of premature packing failure.

Measure and Record

- Document the shaft or stem diameter, stuffing box bore and depth, and when using lantern rings, distance of the port to the bottom of the stuffing box.

Select Packing Material

- Use the correct cross-section of packing or die-formed rings, making certain that the packing is as specified by the packing manufacturer.
- Calculate packing cross section and number of rings needed from recorded measurements.
- Examine packing to be sure it is free from defects.
- Refer to any special installation instructions from packing manufacturer.
- Ensure cleanliness of equipment and packing before proceeding. See Figure III-12.

Figure III-12
Roll Packing for Cutting Different Size Rings

Prepare Rings

- Wind packing around a sized mandrel, or use a calibrated packing ring cutter.
- When using coil or spiral packing, always cut the packing into separate rings. Never wind a coil of packing into a stuffing box. Rings can be cut with butt (square) or skive (or diagonal) joints, depending on the method used.
- Cut packing cleanly, per instructions from packing manufacturer or plant engineering department. See Figure III-13.
- Using the shaft or stem to check for proper sizing, cut one ring at a time and according to instructions from the packing manufacturer to ensure that the rings are sized precisely to fit the shaft or stem.
- Never roll or hammer fiber packing, as that will damage the fiber and can cause premature failure.

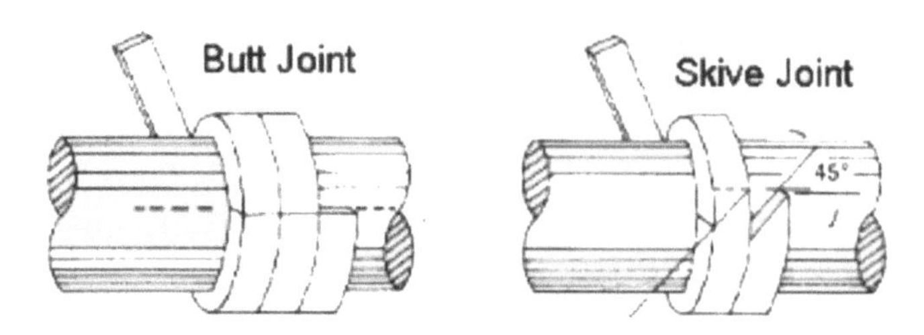

Figure III-13
Two Recommended Joint Types

Install Packing

- Carefully install one ring of the packing at a time.
- Twist each ring over the shaft or stem.
- Ensure each ring is seated fully in the stuffing box prior to installing the next ring.
- Stagger the joints of subsequent rings a minimum of 90 degrees.
- After the last ring is installed, draw the gland up evenly with the nuts finger-tight.
- Check the lantern ring, if used, for correct positioning relative to the port.
- Make sure that the shaft and stem turns freely.

Note: The mechanical pressure curve in Figure III-14 shows eight packing rings. The first five rings do the majority of the sealing. The bottom three rings do little sealing, but are needed to fill the available space. The advantage of using fewer rings is less rod wear.

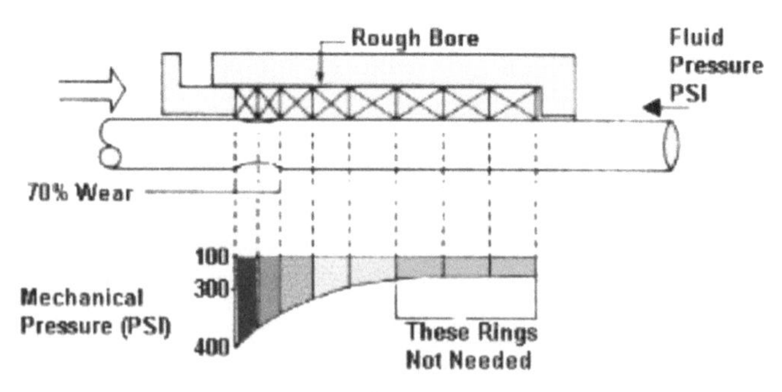

Figure III-14
Mechanical Pressure Curve

Adjust Packing (Pumps)

- Take up the gland nuts until finger-tight.
- Start the pump and tighten the gland nuts allowing liberal leakage. When starting a new pump, allow the packing to leak freely. Excessive leakage during the first hour of operation will result in a better packing job over a longer period of time. Take up gradually on the gland as the packing seats, until leakage is reduced to a tolerable level, preferably 8-10 drops per minute per inch of shaft diameter.
- Reduce the leakage gradually by tightening the gland nuts slowly until leakage reaches an acceptable level. See Figure III-15.

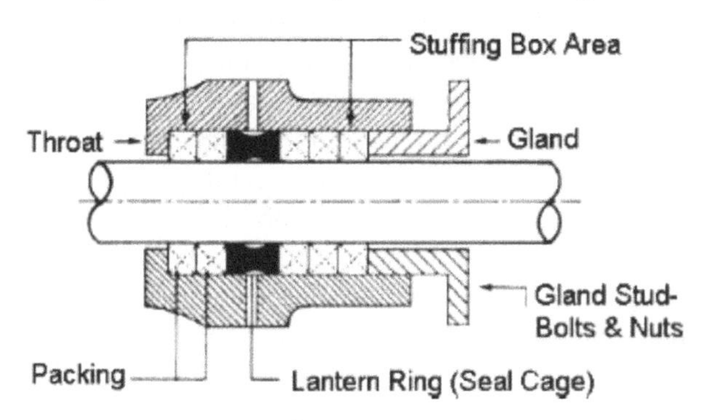

Figure III-15
Typical Pump Shaft Stuffing Box

- If leakage stops abruptly, back off the gland and readjust to prevent the packing from overheating.
- Allow sufficient time between adjustments for the leak rate to stabilize.

Adjust Packing (Valves)

- Consult the packing manufacturer for guidance on torque specifications or percent of compression.
- Tighten gland nuts in multiple steps:

  Step 1—Torque gland bolts to approximately 30% of full torque or appropriate compression percentage.

  Step 2—Cycle the valve a number of times and apply full torque while valve is in closed stroke position.

  Step 3—Repeat Step 2 three or four times.

PROTECTIVE PLANT COATINGS

Facilities are subject to environments and uses that accelerate their natural deterioration and require costly repairs and maintenance. Metals corrode in aggressive soil, industrial or chemical atmospheres, or immersion environments; woods swell, warp, and crack during weathering; concrete and masonry structures crack and spall in severe environments; and organic polymeric materials suffer degradation in sunlight. Paints and other surface coatings provide protection of material surfaces, as well as being decorative and eye-catching. Generally, paint and/or coating types and means of application are dependent upon what function the coating must perform. A coating is a covering that is applied to the surface of an object and is usually referred to as the substrate. In many cases coatings are applied to improve surface properties of the substrate, such as appearance, adhesion, wet-ability, corrosion resistance, wear resistance, and scratch resistance. The types of ingredients within a coating may be obtained from material safety data sheets (MSDS), technical data sheets, or coating vendor technical staff. Coatings are made up of two categories of constituents, solids and solvents. Those categories can be further subdivided as shown in Figure II-16.

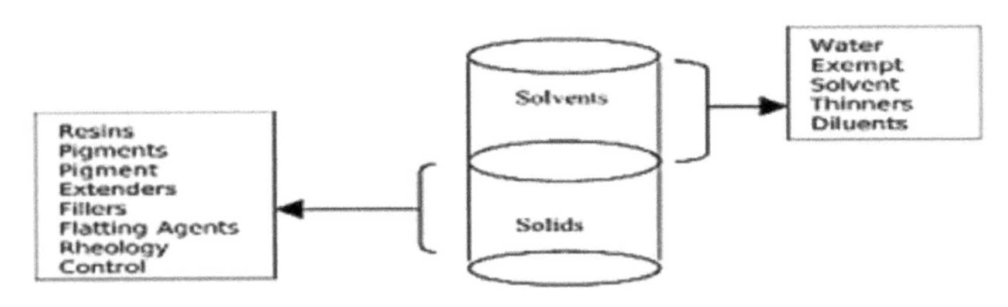

Figure III-16
The Two Coating Constituents

General steps for painting and coating applications typically include substrate surface preparation, application of the coating, and drying of the coating. Coatings protect metals from corrosion by interfering with one or more of the conditions necessary for corrosion to occur. The three protective mechanisms of coatings are barrier, inhibitors and cathodic. Special coatings offer a number of advantages. Applied correctly, special coatings offer excellent finish and gloss to surfaces. Characteristics of special coatings include water repellency, heat resistance, exceptional adhesion, affinity for concrete, preservation and protection. See Figure III-17.

Figure III-17
Coatings Come in a Wide Variety

Coating Constituents

Coatings consist of resins, pigments, solvents and additives. Particular types of coatings we apply will have varying amounts of each of these constituents. Resins or binders hold all paint constituents together and enable them to cure into a thin plastic film. Resins are made of polymers which are chosen based on physical and chemical properties desired in the finished product. Acrylics produce a shiny, hard finish with good chemical and weather resistance. Alkyds are relatively low in cost and because of their versatility are considered general-purpose paint. Epoxies provide excellent water resistance and superior chemical resistance, but do lose their gloss from ultraviolet light. Urethanes combine high gloss and flexibility with chemical stain resistance and demonstrate excellent water resistance.

Applications

There are innumerable ways and places for utilizing coatings in the physical plant, and there are myriad types. Dry film lubricants are an attractive alternative to fluid lubricants for minimizing friction and preventing seizing and galling, especially in high- or low-temperature environments where fluids may freeze or vaporize. Non-stick (fluoro-polymer) coatings are popular for their endless variety of industrial uses. Corrosion-resistant coatings protect metal components against degradation due to moisture, salt spray, oxidation or exposure to a variety of environmental or industrial chemicals. Chemical resistant coatings offer a wide variety of valuable attributes, including abrasion resistance, durability, lubricating properties and easy clean-up. Baked-on coatings minimize friction while offering protection from chemicals, wear, abrasion and corrosion.

Causes of Failure

The majority of paint and coating-related failures can be attributed to six primary causes:

Improper Surface Preparation

The substrate surface is not adequately prepared for the coating that is to be applied. This may include cleaning, chemical pretreatment or surface roughing.

Improper Coating Selection

Either the paint or coating selected is not suitable for the intended service environment, or it is not compatible with the substrate surface.

Improper Application

This can be a problem with either shop-applied or field-applied coatings and occurs when the required specifications or parameters for the application are not met.

Improper Drying, Curing and Over Coating Times

Again, this problem relates to a lack of conformance to the required specifications or parameters.

Lack of Protection against Water and Aqueous Systems

This is a particularly serious problem with aqueous systems containing corrosive compounds such as chlorides.

Mechanical Damage

This damage is caused by improper handling of the painted or coated substrate, resulting in a breach in the paint or coating.

Note: Coating thickness measurement is important. On metal surfaces, defective coatings can lead to rust.

Repair

Surfaces and substrates require regular inspection and repair, as necessary. At least annually, each coated structure should be inspected for deterioration of both the substrates and their coatings. Both the types and the extent of deterioration should be noted, and the generic type of the finish coat should be determined if not already known. An estimate should also be made as to when structural and coating repairs should be made to prevent more serious damage. This is necessary to determine the general scope of the work. There are four different approaches to maintaining an existing coating in an acceptable condition:

- Cleaning only to restore to an acceptable condition. This may be accomplished by pressure washing or steam cleaning.
- Spot repair (priming and top-coating) of areas with localized damage. This should be done before the damage becomes more extensive.

Figure III-18
Sandblasting Booth

- Localized spot repair plus complete refinishing with topcoat only. This should be done when localized repair only would produce an unacceptable patchy finish.
- Full and complete removal of existing paint and total repainting. This should be done when the damage is so extensive that other types are impractical or uneconomical.

Repair of exterior coatings may not be warranted with the first appearance of weathering, but deterioration should not proceed to the point that damage occurs to the substrate, or more costly surface preparation or application techniques become necessary. The plan for maintenance includes selection of the surface preparation, application, and inspection methods and the materials to be used. Organic materials deteriorate and fail with time. Failure analysis does not concern itself with that type of deterioration. It is defined as an investigation to determine the cause of premature deterioration of coating.

TECHNICAL OPERATIONS LIBRARY

Beyond skill sets, information is an essential element of effective maintenance work. Technical libraries house information essential to the care and repair of a facility's building support equipment, and provide the basis for building historical data files on equipment, systems and their operation. These repositories of technical information allow mechanics, technicians and engineers quick and easy access to data they require in understanding and completing their work. Having the right settings, parameters, part numbers, diagrams, and drawings simplifies maintenance work.

Content

The primary (anchor) documents to be found in the library are the original equipment manufacturers manuals (OEM) which form the backbone of the information repository. From them is derived information regarding:

- Life cycle expectations
- Installation and operating instructions
- Standard repair times
- Troubleshooting guidelines

- Parts numbers
- Safety procedures
- Critical spares and materials

They also serve as fodder for the creation of (equipment specific) preventive maintenance and repair work plans, as well as standard operating procedures (SOPs). Other items included in the tech library may be:

- Exploded parts diagrams
- Wiring diagrams (electrical single lines)
- Catalogs, brochures and individual cut sheets
- Copies of original site and plant drawings and specifications
- (Red lined) as-built drawings
- Vendors' catalogs, contact information and training
- Warrantees and guarantees
- Equipment and systems textbooks
- Engineering manuals

Configuration and Utilization

Information comes in many forms—hardcopy, physical (as in a display cabinet where worn and broken parts, burned bearings and other failed items can be viewed, picked up and handled for learning and orientation purposes). Technical libraries are becoming increasingly virtual (computerized) and electronic—to the extent that information is accessed via organizational intranets and the internet. It is especially handy for maintenance technicians to have access to a computer and printer. The advantage is that the capabilities are available around the clock, 365 days a year. Personnel can then use the local network or the internet to look up such things as repair history jackets, computerized maintenance management (CMMS) documents and OEM (original equipment manufacturer) manual data. A technical library can be a great aid in performing economic analysis of work task lists, establishing justifications for and coordinating shutdowns, creating protocols for analysis of data and other paperwork exercises, such as:

- Repair versus replace decisions
- Repair history trending
- Applicability of repair technologies
- Continuous improvement methodologies
- Time standard development

- Testing and certification reports
- Maintenance records and consumables used
- Occupant requests and service level agreements (SLAs)

Oversight

The basis for successful scheduling, monitoring, and troubleshooting of equipment is comprehensive and up-to-date documentation. There should be a single point of contact for all library content, responsible for:

- Keeping the information organized and current (substituting revisions and discarding materials as appropriate).

- Maintaining all materials in an environmentally friendly atmosphere (temperature, humidity, fire...).

- Signing out removed materials and providing a receipt upon its return. See Table III-5.

THREADED FASTENERS

Screws and bolts (threaded fasteners) hold parts of a structure together by transferring load from one component to another. They come in a wide variety of types (forms) and are utilized in myriad applications. See Figure III-19. Their designs and functions determine into which categories these different types of fasteners fall. You can usually tell their intended uses by their names. "Wood screws" are specifically made for use in wood, "sheet metal screws" are designed to be used in sheet metal and "machine screws" (with standardized threads) are used for connecting machine parts together. Screws provide more strength and holding power than nails, can easily be removed and are available with different coatings to deter rust. They are manufactured with four basic heads and different kinds of slots:

- Flathead—countersunk into the material so the head of the screw is flush with the surface.
- Oval-head screws—partially countersunk, with about half the screw head above the surface.

- Roundhead screws—(not countersunk); the entire screw head lies above the surface.
- Fillister-head screws—above the surface to keep the screwdriver from causing damage.

Table III-5
Original Equipment Manufacturer Manuals Log

			SECTION / CHAPTER						ASSET	
ID #	MFR	DEVICE	Safety	Dynamics	Install	Maint	Oper	Trouble	FAC #	Location
OEM1	Quincy Corporation	Air comp	x		x	x	x	x	00647COMP RESSOR0000 1	14A Basement
	Quincy Corporation	Air comp	x		x	x	x	x	00647COMP RESSOR0000 2	14A Basement
OEM2	Larkin Heatcraft	AC Comp			x	x	x	x	XXXXX	14A N. Penthouse
	Larkin Heatcraft	AC Comp			x	x	x	x	XXXXX	14A N. Penthouse
OEM3	Quincy	Air Dryer	x		x	x	x	x	00647DRYER 00001	14A Basement
OEM4										

(Continued)

Table III-5 (*Continued*)
Original Equipment Manufacturer Manuals Log

OEM4	Hankinson	Air Dryer	x		x	x	x	x	09281AIRDR YER00001	14B N. Penthouse
	Hankinson	Air Dryer	x		x	x	x	x	09282AIRDR YER00001	14B N. Penthouse
	Hankinson									

- Signing out removed materials and providing a receipt upon its return.

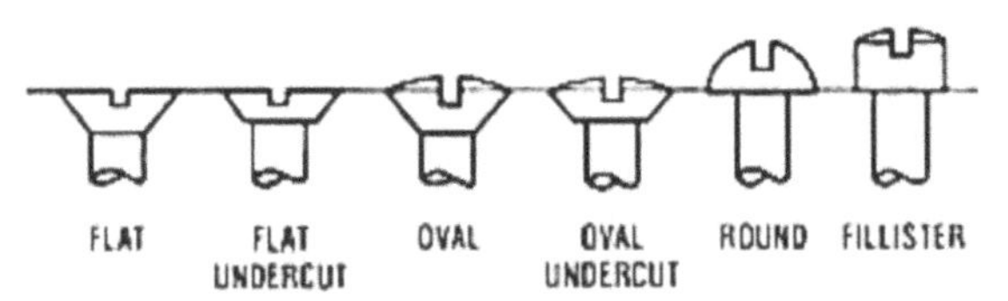

Figure III-19
Screw Head Assortment

Most screws have slot heads and are driven with slotted, or standard, screwdrivers. Phillips-head screws have crossed slots and are driven with Phillips screwdrivers. Screws are measured in both length and diameter at the shank (which is designated by gauge number from 0 to 24). Length is measured in inches. The length of a screw is important because at least half the length of the screw should extend into the base material. See Figure III-20.

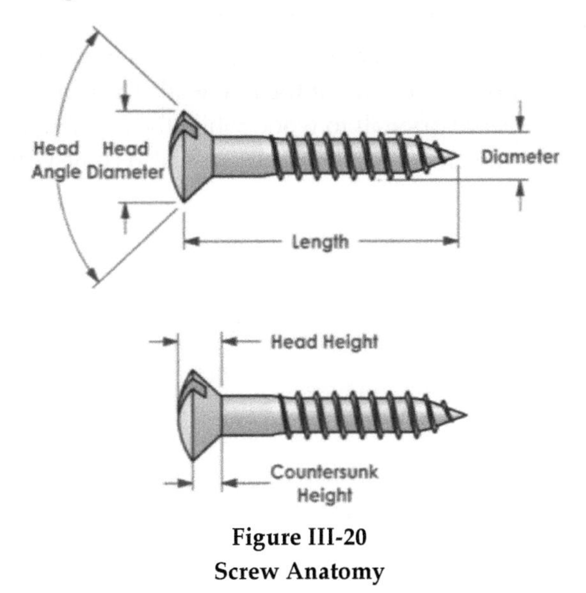

Figure III-20
Screw Anatomy

Note: To prevent screws from splitting the material, pilot holes must be made with a drill before the screws are driven.

Screw Types

Wood screws are usually made of steel, although brass, nickel, bronze, and copper screws should be used if there is potential for corrosion. They have a coarser pitch (few threads per inch) than sheet metal or machine screws, and often have an unthreaded shank. The thread less shank allows the top piece of wood to be pulled flush against the under piece without getting caught on the threads. **Note**: Some wood screws are tapered from tip to head. See Figure III-21.

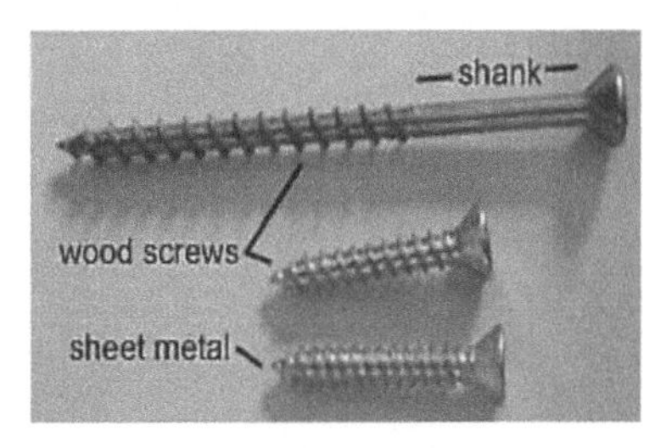

Figure III-21
Wood and Sheet Metal Screws

Sheet Metal Screws—Usually threaded all the way to their head, these will work in wood, but wood screws shouldn't be used in metal. Most of these screws are self-tapping in that they only require a pre-drilled hole (pre-drill sizes), but some come with self-drilling or self-tapping tips.

Drywall Screws—The head-to-shaft junction is more curved than in a wood screw to prevent tearing of the dry-wall. The coarse thread version is meant to secure drywall to wood while the fine thread version is for attachment to metal studs. **Note**: These can also come with self-drilling tips.

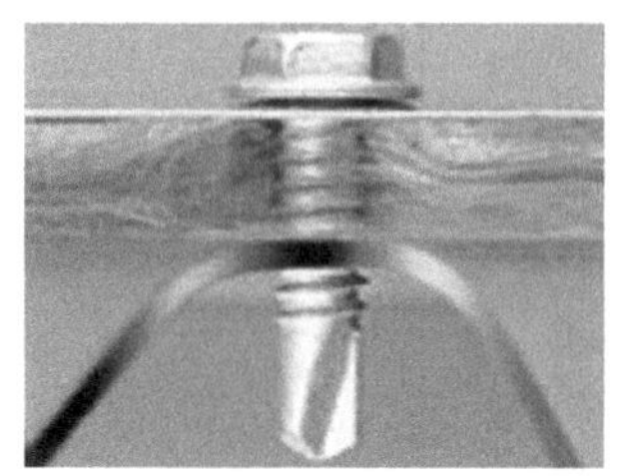

Figure III-22
Drywall Screw

Machine Screws—blunt-ended screws used to fasten metal parts together that are commonly made of steel or brass. They are also made with coatings—brass, copper, nickel, zinc, cadmium, and galvanized—to help deter rust. Machine screws are manufactured with each of the four basic types of heads—flathead, oval-head, roundhead and fillister-head—and with both plain and Phillips head slots. They are available in gauges 2 to 12 and diameters from 1/4 inch to 1/2 inch and in lengths from 1/4 inch to 3 inches. See Figure III-23.

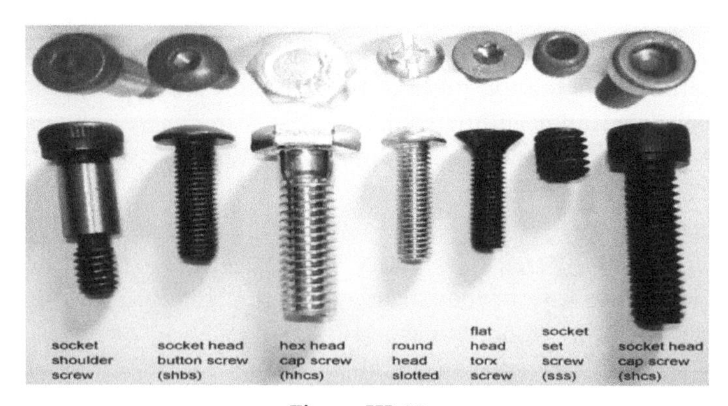

Figure III-23
Machine Screws

Lag Screws—heavy-duly fasteners (expansion anchors and lag screws) are used for larger jobs and more holding power. They are driven with a wrench and are used primarily for fastening to masonry or wood framing. The anchors are inserted into holes drilled in the masonry, and the lag screws are driven firmly into the anchors. See Figure III-24.

Figure III-24
Lag Screw

A hex head cap screw and hex bolt may look identical; so... what's the difference between a bolt and a screw? The most generally accepted "rule-of-thumb" definition (in terms of how they are installed) is as follows: If you turn a head, it's a screw; if you turn a nut, it's a bolt.

Bolt Types

A "bolt" is a screw with a flat head and no point, for fastening things together. Bolts are used with nuts and often with washers. The three basic types are carriage bolts, stove bolts and machine bolts. Other types include the masonry bolt and anchor, toggle bolt, and expansion bolt, which are used to distribute weight when fastening something to a hollow wall. Machine bolts are manufactured in two gauges, fine-threaded and coarse. Carriage and stove bolts are coarse threaded. Bolt size is measured by shank diameter and by threads per inch, expressed as diameter by threads (for example, 1/4 X 20). Carriage bolts are available up to 10 inches long, stove bolts up to 6 inches, and machine bolts up to 30 inches.

Carriage Bolts—Carriage bolts have a round head with a square collar and are tightened into place with a nut and wrench. The collar tits into pre-bored hole or twists into the material, preventing the bolt from turning as the nut is tightened. Carriage bolts are coarse-threaded and are available in diameters from 3/16 to 3/4 inch and lengths from 1/2 inch to 10 inches. See Figure III-25.

Figure III-25
Carriage Bolts

Stove Bolts—can be used for almost any fastening job and are available in a wide range of sizes, have a slotted head—flat, oval, or round, like screws—and are driven with a screwdriver or tightened into place with a nut and wrench. Most stove bolts are completely threaded, but the larger ones may have a smooth shank near the bolt head. Stove bolts are coarse-threaded and are available in diameters from 5/32 to 1/2 inch and lengths from 3/8 inch to 6 inches. See Figure III-26.

Figure III-26
Stove Bolt

Machine Bolts—are manufactured in a number of sizes, have either a square head or a hexagonal head, are wrench driven and fastened with square nuts or hex nuts. The bolt diameter increases with length. They are either coarse-threaded or line-threaded and come in diameters from 1/4 inch to 2 inches and lengths from 1/4 inch to 30 inches. See Figure III-27.

Material Selection

Depending on the application and the environment in which threaded fasteners are used, they are manufactured with a vast array

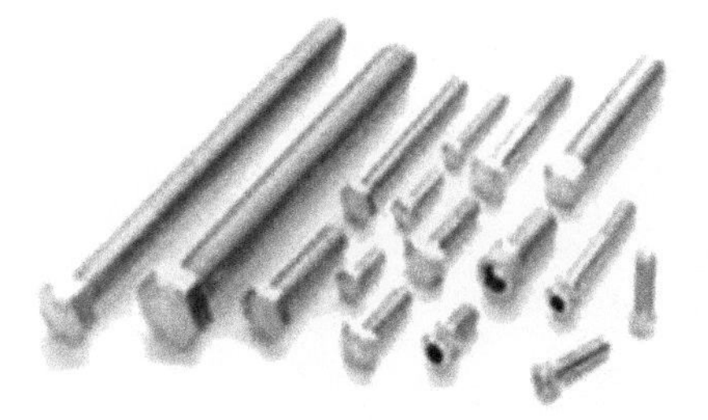

Figure III-27
Machine Bolts

of different materials, platings and coatings. See Figures III-28 and III-29. Material selection is influenced by required strength, temperature, resistance to corrosion, joint materials, and cost. When placed in contact along with an electrolyte (like humid air), certain metals corrode each other (galvanic action). It happens when the metals are substantially electro-chemically different from one another (such as at copper and aluminum electrical connections). Some of the more common materials used to avoid that problem include:

Zinc Plating—(the most common covering) is utilized for outdoor conditions, due to steel's tendency to rust.

Black Oxide—provides very mild protection against corrosion and usually has an oil film added for additional protection. It is most commonly applied to socket head cap and other machine screws.

Hot-Dipped Galvanized—provides the best (common) protection next to stainless steel.

Stainless Steel and Aluminum—are inherently resistant to corrosion because they form a tough oxide layer when exposed to oxygen. The strength of stainless is much less than alloy steel, and even less so for aluminum.

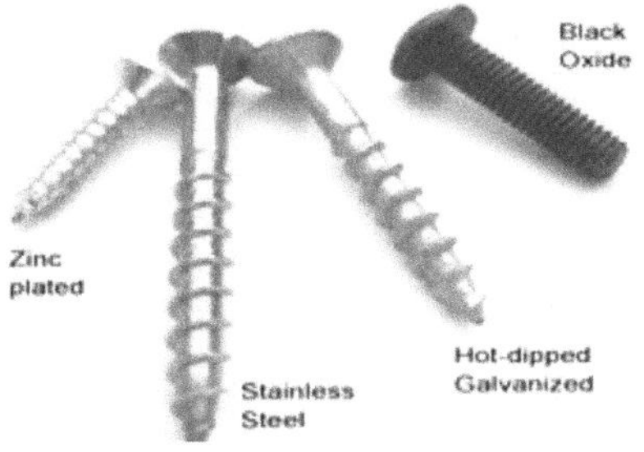

Figure III-28
Different Coatings

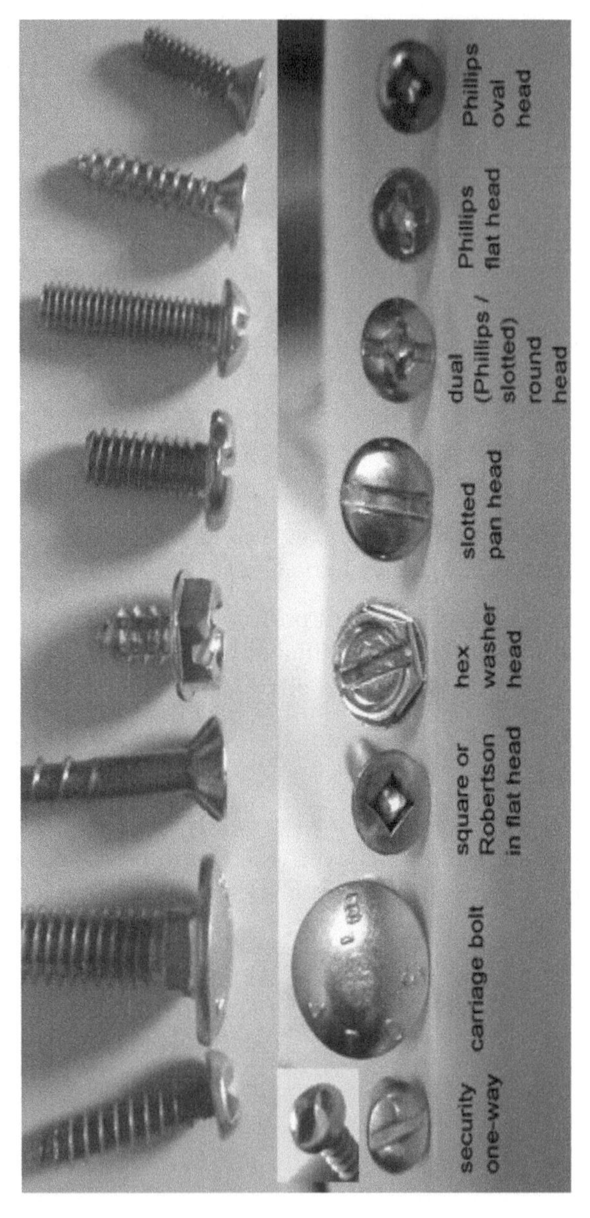

Figure III-29
Screw and Bolt (Head) Variations

Washers

Washers help to evenly distribute the load, prevent screw and bolt heads from digging into the joined materials and lock threaded fasteners in place. Without them, it's more likely that the screw or bolt will compress higher spots over time and come loose. There are two different standards commonly used in the manufacture of washers. The first is USS (United States Standard); the second is SAE (Society of Automotive Engineers). USS standard washers are the most common type, having a larger outside diameter to help distribute the forces applied when tightening. SAE standard washers are very similar to the USS except that it has a tighter fit to the bolt and has a smaller outside diameter. This picture shows a comparison of an SAE and USS washer. Note that the USS washer has a larger diameter and a looser fit around the fastener. It is also a little thicker. There is an endless variety of washers (by composition) available depending on the particular applications they are used in. The most common types include:

Fender washers—have an over-sized outside diameter but are thinner and usually made with a cheaper material (as opposed to USS).

Lock washers—are used to help keep the nut from loosening from the bolt. When tightened, the lock washer "locks" the nut in place. They come in two varieties, external—having external teeth that help lock the nut in place, and internal—having internal teeth that help lock the nut in place. See Figure III-30.

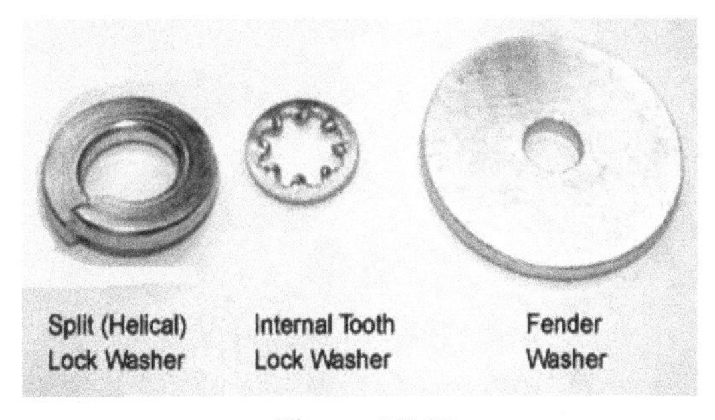

Figure III-30
Washer Types

Split lock washer—has two features that prevent loosening, a spring action and an edge that digs into the screw on reversal of the screw or bolt. According to the *Handbook of Bolts and Bolted Joints,* the lock washer undergoes additional deformation after it flattens with a spring rate more comparable to that of the bolt. This extra springiness is helpful for preventing fatigue failure, but it's unlikely that it helps prevent loosening. These washers are probably most effective in joints where the recommended tightness cannot be achieved, such as soft metal, plastic or wood joints. In these cases, the washer would likely not be entirely flat and would indeed dig into the screw surfaces.

Toothed washer—has small teeth that dig into adjacent screw and joint material. They seem to be more effective than split lock washers, but will (and must) cause damage to adjacent surfaces, which may affect repeated installation.

Belleville washers—are cone shaped washers used more as a precision spring than a locking device. They can be stacked to increase their combined spring rate. Their spring rates are substantially higher than split lock washers. They may provide some protection in high vibration or temperature changes. Wavy washers have a similar purpose. See Figure III-31.

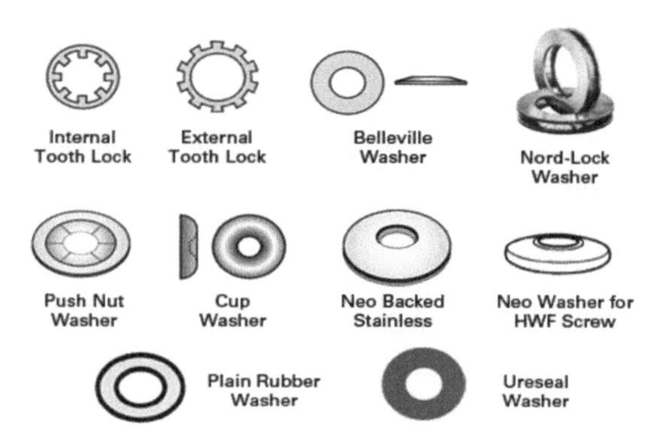

Figure III-31
Belleville Type Washers

Failures

A survey completed in the United States by automobile service managers indicated that 23% of all service problems were traced to loose fasteners, with even 12% of new cars being found to have fasteners loose. Whereas loose fasteners account for myriad machinery and equipment failures, actual physical failure of screws and bolts themselves is also a problem.

Shear stress is the sheer force transferred, divided by the cross sectional area, which is generally a circle. See Figure III-32.

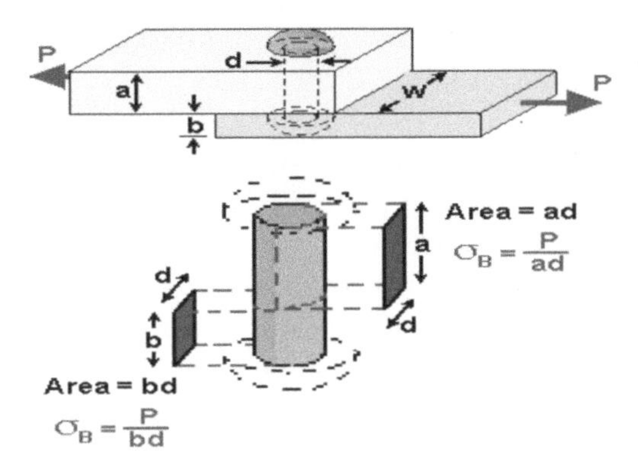

Figure III-32
Sheer Stress

Bearing stress is caused by one component acting directly on another.

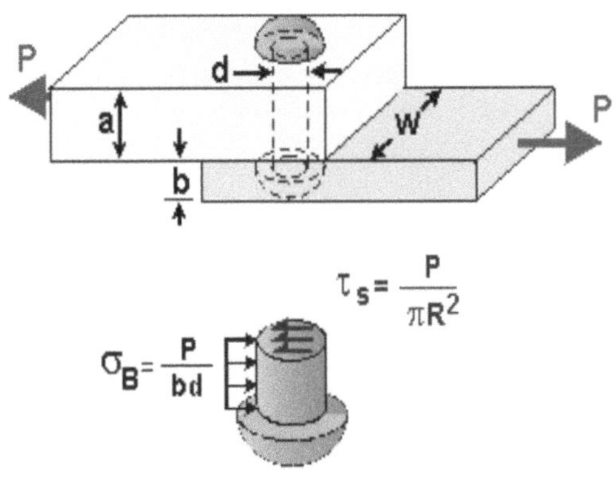

Figure III-33
Bearing Stress

Note: Locking the male-to-female thread by means of adhesive gives fastened assemblies excellent resistance to loosening due to vibration. Adhesives may be applied in liquid form at the assembly stage, or the threads may come coated previously with an anaerobic adhesive which cures when the parts have been assembled. Most thread locking adhesives tend to display a thread friction coefficient which is higher than what is normally present in the threads.

VALVE EXERCISING AND LUBRICATION

Every fluid system has valves—devices that regulate, stop, or start the flow of water in the distribution lines. Being able to operate these valves at a moment's notice is extremely important. In an emergency, sections of a distribution system may need to be shut down without delay. However, if a valve is not used over a period of time it can seize up from corrosion and get stuck, making the valve inoperable. The continued operating integrity of valves of all types (such as distribution and transmission valves, air valves, relief valves and blow-offs), requires regular inspection and operation. A systematic program is required to locate and exercise water valves to ensure that they function/operate properly. Each valve should be operated through a full cycle and returned to its normal position on a schedule that is designed to prevent a buildup of tuberculation or other deposits that could render the valve inoperable or prevent a tight shut-off. This primer outlines key elements of a comprehensive valve exercising program.

Rationale

An effective valve exercising program is essential to improve customer service, ensure mission capability, ensure distribution system reliability, develop predictive maintenance programs, determine capital improvement budgeting, develop loss trend analysis, ensure system isolation capability, and ensure quality control. Knowing where the valves are will help with obtaining detailed information on the valve type and size, as well as locating the rest of the distribution system, which is often an issue with older utilities. Some of the benefits of fully operational valves are given below.

Benefits

- Confidence the valves will last much longer
- Significant reductions in labor costs for repairing emergency leaks

- Enable the isolation of portions of the system (system partitioning)
- Proactive diagnosis and repair of bad valves before you need them.
- Reduced system losses and property damage by being through rapid system isolation
- Minimize service interruptions
- Exercised valves can work for a lifetime
- Reduced labor costs by being able to find and use the "key" valves in an emergency
- Improved valve reliability (less frequent repairs)
- Identification of critical valves in a distribution system
- Development of a trend analysis capability
- Measurement and documentation of valve operation

Key Plan Components

To set up a program, find and document the valve's location and take a digital picture showing the valve and surrounding area, operate the valve until it operates freely (with little resistance). Operate it through at least one full cycle, and several if needed. (A more detailed discussion on the actual exercising is found below). Maintain detailed records for each valve. This includes mapping locations on as-built drawings or roadmaps and maintaining both electronic and hard copies. (Record keeping is discussed in more detail below). Schedule and perform needed repairs. Valves are sometimes broken during the exercising program because they have not previously been used. (Fixing broken valves in a timely manner is very important). Repeat these steps on a routine basis. Experts recommend exercising a system's valves annually if possible, or at least once every two years. Some valves will need to have a different schedule than others based on their location or unusual operating conditions. For exterior located valves, it is usually a good idea to perform the exercising program during moderate weather conditions.

Program Initiation

The first step in a comprehensive valve-exercising program is to prioritize the valves. The most important valves are usually those near critical customers (operations). Other important factors include knowing the amount of flow through the valves, their age and location. Starting the program consists of the following actions:

- Gather all information on your distribution system
- Start small—one valve at a time
- Start with known valves or critical valves
- Expand to locate all valves
- Record information
    - Location
    - Valve size
    - Number of turns
    - Depth to valve nut
    - Valve head—square vs. wheel valve
    - Open direction—left vs. right
    - Date exercised

Maintenance Required/Completed

Develop a spreadsheet or report with all the valve numbers, locations and sizes to create a maintenance record showing the date the valve was exercised and what condition the valve is in and develop a plan to continue the program.

Inspection and Lubrication

- Inspect packing gland (tighten and lubricate as necessary).
- Check for correct positioning and operation.
- Check for leaking seals and gaskets.
- Adjust operator linkages and limit switches on control valves.
- Exercise and grease stems on OS&Y (outside screw and yoke) valves.
- Wipe valve operator rods clean and apply coat of light oil.
- Lubricate valve stems, and exercise valves to verify operation and distribute lubricant.
- Check sealed valves to verify valve is sealed in open position and seal is unbroken week.
- Check locked valves and valves equipped with electric tamper

switches to verify valve is open, and lock is not broken or tamper switch is not damaged.

Exercising Valves

If valves haven't been used in some time (or ever), you will encounter difficulties during the exercise program. When exercising valves, take ample time and never force the valve! Use the lowest torque (turning force or rational force) setting possible. Avoid using a cheater bar (a handle extension that allows for greater torque). A cheater bar should only be used in emergencies. If and when the valve turns freely, turn it slowly to avoid water hammer. If a valve is opened or closed too rapidly, the line may be subjected to extreme pressure changes and it may burst. Listen closely. Sometimes a change in flow can be detected when opening a valve. Debris can be stirred up during valve exercising, notify affected personnel before starting the process. (this will keep the complaint calls down). Consider conducting the flushing program at the same time as exercising the valves. **Note**: Always count the turns to open and close; they should match.

The American Water Works Association (AWWA) provides these guidelines:

- Verify the direction for turning the valve to the closed and open positions.
- Assume valve is in the full open position.
- Begin closing valve slowly, increasing torque as necessary to achieve movement (without exceeding the pre-determined maximum torque).
- Count the number of turns necessary to achieve the full open position.
- Begin opening valve slowly, increasing torque as necessary to achieve movement (without exceeding the pre-determined maximum torque).
- Count the number of turns necessary to achieve the full closed position.
- Repeat the close/open cycle a minimum of three times, or until the number of turns necessary to open or close the valve does not change.

- Record the number of turns, cycles, and maximum torque applied.

(To properly close a valve:)

- Begin with a steady amount of torque in the direction necessary to close the valve, moving through five to 10 rotations.
- Reverse for two or three rotations.
- Reverse again and rotate five to 10 more turns in the closing direction.
- Repeat this procedure until full closure is attained.
- Once the valve is fully closed, it should be opened a few turns so that high-velocity water flowing under the gates can move the remainder of the sediment downstream with more force and clear the bottom part of the valve body for seating.
- Fully close the valve again.

The reason for this cautious approach is that debris and sediment often build up on the gates, stem, and slides. If this material is compacted while the valve is being closed, the torque required to close the valve continues to build as the material is loaded. If the procedure described above is used, the stem and other parts are scrubbed by the series of back-and-forth motions. Remember that valve manufacturers have detailed operation and maintenance procedures for each of the various types of valves. Some valves have a seating where a resilient coating meets stainless steel. Other valves have actuators isolated from the water flow, meaning that some of the mechanical parts are not subject to as much corrosion and, therefore, may need less exercise. When in doubt, follow the manufacturers' guidelines.

WATER SYSTEMS MAINTENANCE

Water systems are the life blood of facility operations. Properly maintained water distribution systems are important for ensuring the delivery of high quality water to customers, extend equipment life-cycles and minimize problems related to minor or major equipment failures. Maintaining the integrity of these important systems is paramount for preventing contamination events, ensuring continuous service during emergencies and minimizing property damage. Water distribution systems consist of the items shown in Figure III-34.

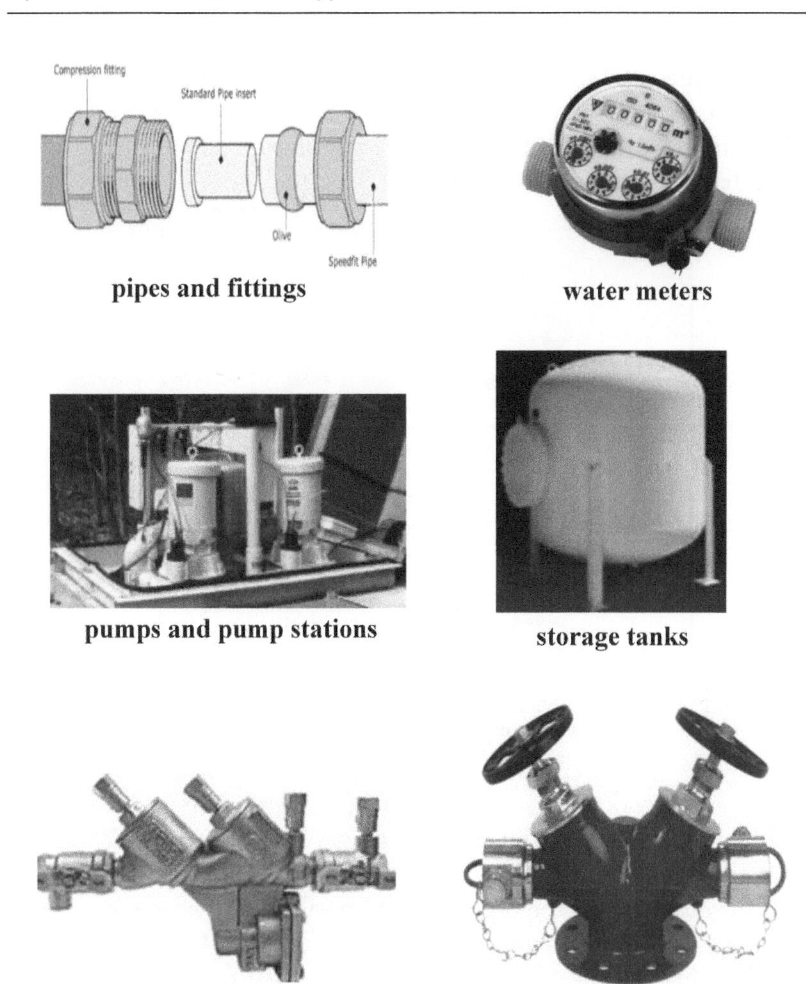

Figure III-34
Key Elements of Water Distribution Systems

Pipe Fitting Basics

Pipe fitting is the installation or repair of pipes that transport liquids or gases. Proper pipe fitting involves selecting and preparing pipes from a variety of materials including copper, plastic, steel, iron, aluminum and lead. The method of fastening your pipes together will be determined by what type of material you are working with. Plastic PVC pipes require PVC cement that is applied to both the ends of the pipe and inside the fittings. Apply the PVC cement using a PVC cement brush, and reassemble the pipe and fittings. Twist the fittings 30 to 45 degrees to ensure complete cement coverage, and wipe away any excess with a damp rag. It typically takes 1 to 2 minutes for a PVC fitting to set up and dry. See Figure III-35.

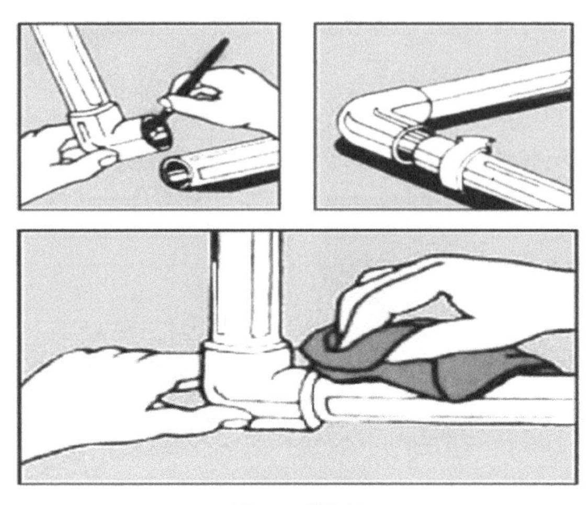

Figure III-35
PVC Plastic Pipe

Fastening copper pipes and fittings together involves a technique called "sweating." Clean the end of the pipe with a fine wire brush, as well as the inside of the fitting. Use a paint brush and apply plumbing flux to the outer end of the pipe and the interior of the fitting. Reassemble the pipe and fittings and using a propane torch, begin to directly heat the fitting. As the flux melts, the copper will get shiny, then moments later it will dull and begin to lightly smoke. At this time, remove the flame and touch lead-free plumbing solder to the joint. The solder will work its way into the joint by capillary action and create a strong, watertight joint. See Figure III-36.

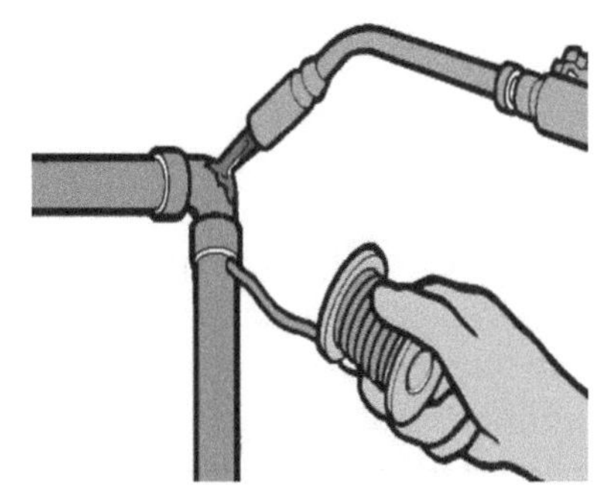

Figure III-36
Sweating a Copper Joint

Note: It is important to only attach like materials together, such as copper to copper and plastic to plastic. Joining different types of tubing together is difficult and requires the services of a certified pipe fitter.

Water Meters

Water meters measure the amount of water used by a source. They measure the volume of water that passes a point in a pipe using mechanical, magnetic or electronic devices, and they can be utilized to determine where water leaks may be occurring in water systems. There are two basic types of water meters (including many variations of each). The first is a positive displacement water meter, and the second is a velocity water meter. The simplest mechanical meters measure the displacement of rotary pistons or loosely suspended disks as water flows past. They are called "oscillating piston" meters. Other types of meters use speed of flow (velocity meters) to calculate volume. Large turbine meters are used in high-volume pipes. Meters combining both are called compound meters. Compound meters use a valve mechanism to direct water flow into each part of the meter so readings can be taken of both mechanisms. See Figure III-37.

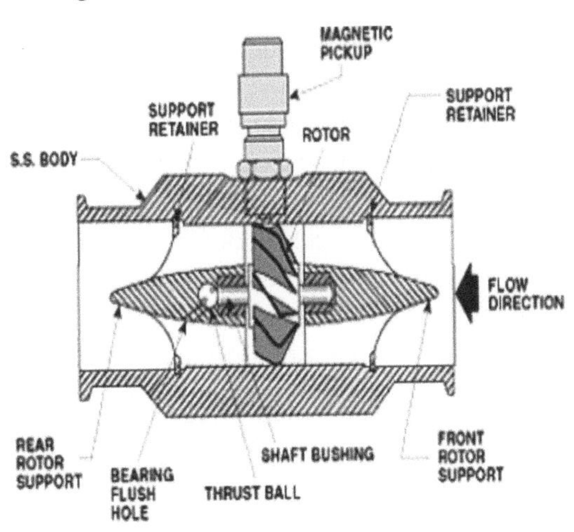

Figure III-37
Turbine Meter

Pump Stations

Package pumping stations provide an efficient and economical way of installing a drainage system. They are suitable for mechanical building services collection and pumping of liquids like surface water, wastewater or sewage from areas where drainage by gravity is not possible. Units are supplied with internal pipe work fitted, pre-assembled ready for installation, after which the submersible pumps and control equipment are fitted. See Figure III-38.

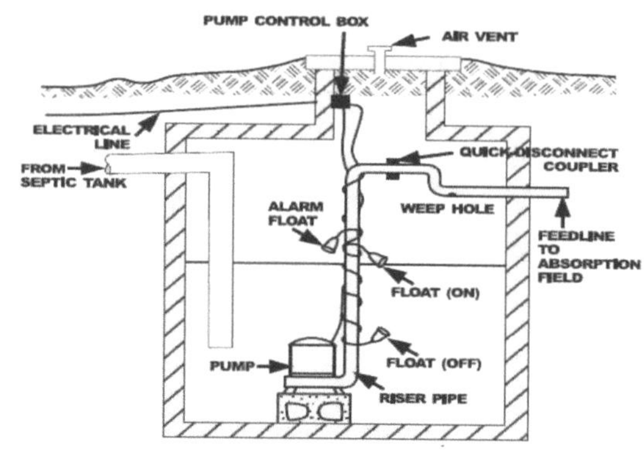

Figure III-38
Pumping Station

Storage Tanks

Water storage tanks hold cold, hot, fresh and brackish waters and can be located above or below ground. Depending on their volume capacity and service, they can be constructed from a variety of materials. Facility managers are mostly confronted with domestic water tanks, sumps and lift stations. Domestic water tanks are vessels used to store fresh (hot or cold) water from which the water is distributed as needed. Hot water tanks are used to store hot water for space heating or domestic use. Hot water tanks may have a built-in gas or oil burner system, or electric immersion heaters, or may use an external heat exchanger to heat water from another energy source. Based on their service, they may or may not be insulated and have myriad accoutrements. See Figure III-39.

Where the water supply has a high content of dissolved minerals such as limestone, heating the water causes the minerals to precipitate in the tank; a water tank may develop leaks due to corrosion after only a few years. In steel tanks dissolved oxygen in the water can accelerate corrosion of the tank and its fittings. A hot water storage tank is wrapped in heat retention insulation to reduce heat loss and energy

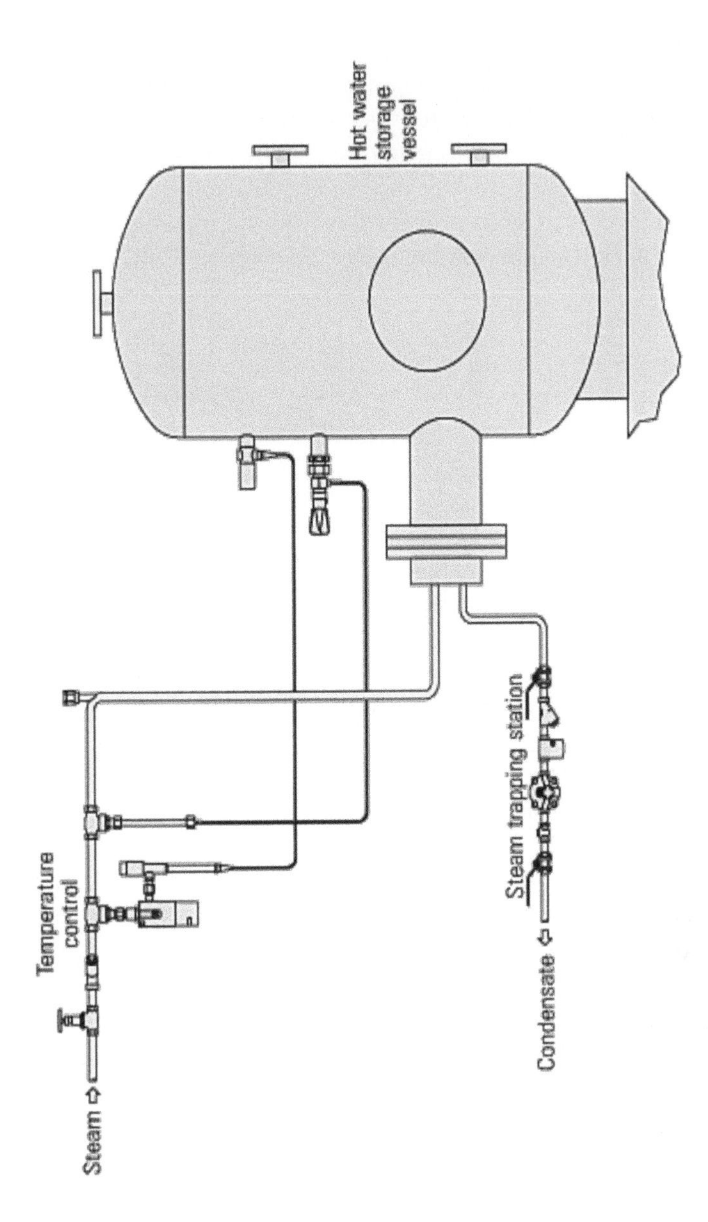

Figure III-39
Hot Water Tank and Apparatus

consumption. A heavily insulated tank can retain heat for days without the application of heat to its contents.

Backflow Preventers

Backflow preventers are mechanisms designed to prevent contaminants from entering potable water systems in the event of back pressure or back siphonage. Back siphonage occurs when a vacuum is created upstream of the backflow device, and water is literally sucked back into the system. The only way to prevent this from happening is the use of some type of backflow prevention device. See Figure III-40.

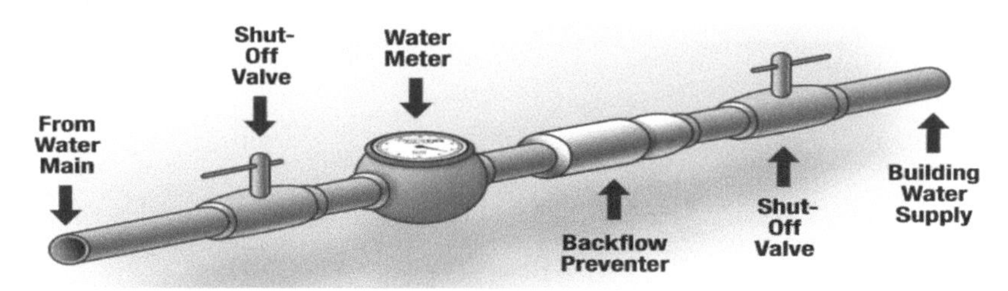

Figure III-40
Backflow Preventer Installed in Building Water Supply Line

> **Note**: Back siphonage can happen when the water supply is stopped due to a water main break or nearby fire hydrant use.

Backflow preventers are a crucial part of any potable water system. To understand what a backflow preventer does, you should understand the following terms:

Backflow

Any unwanted reversal of the flow of liquids, solids or gases in a piping system.

Back Pressure

When the pressure downstream of the backflow devices exceeds the supply pressure.

Cross-connection

Any connection between a potable (drinkable) water system and any system containing non-potable water, pollutants or toxins.

Note: Most areas require the use of backflow preventers on all water systems, but even if your area doesn't have these building codes, they're a mandatory precaution for a health-conscious persons.

Hydrants

Fire hydrants are basically cast iron tubes with appropriate fittings connected to a large underground pipe that carries water. A pentagonal nut on top of the hydrant is used to open an underground valve (the valve is usually several feet underground in an attempt to prevent freezing). When opened, water comes up through the pipe, through the openings and out of the hydrant at a pressure of about 50-80 psi (city water pressure). This pressure is high enough for everyday use, except where booster pumps need to be employed to raise the pressure (such as for use in high-rise buildings). See Figure III-41.

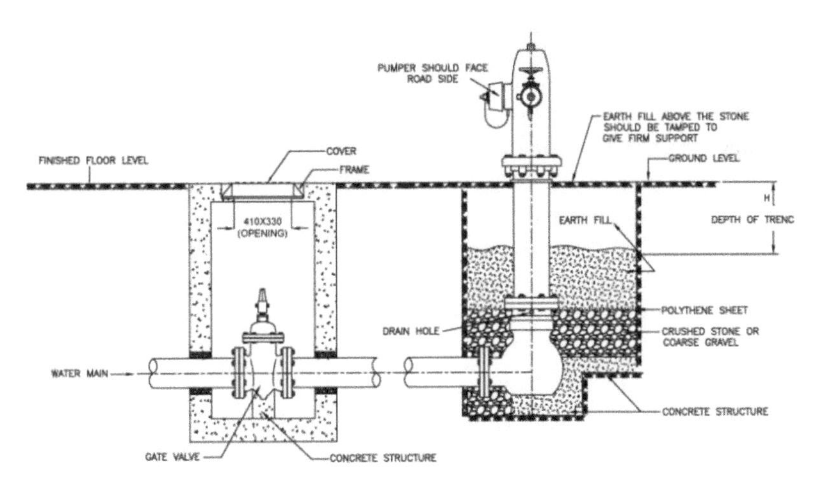

Figure III-41
Fire Hydrant Assemblage

Since fire hydrants are one of the most accessible parts of a water distribution system, they are often used for attaching pressure gauges or loggers or to monitor system water pressure. Automatic flushing devices are often attached to hydrants to maintain chlorination levels in areas of low usage. Hydrants are also used as an easy above-ground access point by leak detection devices to detect and locate leaks from the sound they make. The main challenges of hydrant design—anti-freezing, hydraulic efficiency, ease of repair—are all known and have been dealt with.

> **Note**: In most US areas, contractors who need temporary water may purchase permits to use hydrants. The permit will generally require a hydrant meter, a gate valve and sometimes a clapper valve (if not designed into the hydrant already) to prevent backflow into the hydrant.

System Inspections

Routine inspections of system piping and components should be performed on a regular schedule in accordance with manufacturer's instructions. Frequencies should be based on system experiences and water quality issues and maintenance performed as needed. The intent of the inspections should be to monitor corrosion, check for wear, track leaks and monitor water quality. Corrosion monitoring identifies the need to modify treatment or conduct flushing. Checking for wear (such as in mechanical parts found in pumps and control valves) can extend the useful life of infrastructure components and reduce or avoid unnecessary replacement or operational costs. Tracking leaks can reduce pumping and treatment costs by identifying leaks, breaks, diverted water and inaccurate meters. Monitoring water quality (e.g., pH and temperature) provides information on potential contamination of raw and finished water and determines the effectiveness of the water treatment process.

Maintenance Tasks

Maintenance of system piping and components should be performed on a regular schedule in accordance with manufacturer's instructions. As is the case with inspections, frequencies should be based on system experiences and water quality issues where maintenance is performed as needed. Maintenance actions should include tank maintenance, flushing pipelines, valve exercising, testing for presence of excess bio-films, maintaining operating pressure range and inspecting and flushing hydrants and valves. Storage tank maintenance improves protection against sources of contamination and extends the useful life of the equipment. Testing for the presence of excess bio-films indicates a presence of inadequate chlorine residual, possible high disinfection byproduct levels, and water stagnation. Flushing pipelines removes stagnant water from the pipeline, reduces buildup of bio-films and sediments and restores disinfectant residual. Maintaining operating pressure range of the distribution system reduces the risk of backflow contamination, provides better service to customers reduces damage to infrastructure due to excess pressure and provides adequate fire flow. Valve exercising improves reliability, familiarizes crews with valve locations, identifies inoperable valves, locates obstructed valve boxes, and ensures isolation of the distribution system sections when necessary. Inspecting and flushing hydrants and valves ensures that hydrants and valves are operable, that no water losses occur and that hydrants and valves are not susceptible to tampering.

Section IV

Plant Engineering Principles and Norms

ENERGY ENGINEERING

Electricity is a form of energy, but not all energy is electric. Nowhere is that truth more evident than in our energy intensive physical plants. There is potential energy in our fuel supplies, kinetic energy in our pipes and expended energy in the heat we're constantly transferring, supplying or dissipating with our machines, devices and lighting systems. Energy gains await those with a clever enough plan to conserve that energy within the physical plant. Strictly speaking, energy engineering identifies potential areas of energy waste, assesses their impact on an organization's operational bottom line and prescribes cost-effective alternatives for dealing with them. In maintenance parlance that equates to making choices, such as what fuel to use or how long to leave a system in service. According to Einstein's hypothesis, energy can neither be created nor destroyed. That it can be wasted, he left us on our own to puzzle over. Reducing energy consumption is a lost art. Often we don't establish it as a high enough priority, or provide the proper training for its conservation.

The Building Envelope

A building's shell (or envelope) provides a barrier between the unpredictable atmosphere outside and the controlled environment within. Properly designed and utilized, this barrier, which includes all of the building's structural components (doors, windows, roof, walls, etc.), can selectively filter out or allow in certain amounts of fresh air, light, heat and humidity to temper the building's interior as desired. All of these components have a bearing on the operation of the systems and equipment housed within a building. After recognizing energy's many faces (chemical, mechanical, thermal, etc.), the monitoring and control of that energy is vital to the total operation. A well designed and maintained envelope will: minimize heat losses and gains through proper insulation and control of sunlight; maintain the movement of water vapor and control condensation via good ventilation and humidity control; prevent inappropriate air infiltration and escape through engineered openings; and provide adequate venting of interior gasses with mechanical exhaust systems. Normal maintenance of the structure and its components includes:

- caulking of frames and thresholds
- repair of leaks and cracks
- exchange of dry insulation for saturated
- removal of efflorescent salts
- repair of locks and hardware
- application of protective coatings
- maintenance of roof flashing
- replacement of exterior lighting
- minimization of penetrations
- proper ground level drainage
- replacement and tinting of fenestration
- tuck-pointing of masonry
- proper vent installation
- weather-stripping entryways
- removal of debris from roof drains
- cleaning of windows and curtains
- regular inspections and PMs
- sealing of porous surfaces
- waterproofing areas below ground

General Conservation Measures

Essentially, there are two approaches to reducing energy consumption: One is pro-active, where we take the bull by the horns and wrestle it to the ground; the second is inactive, where we do nothing, and the bull eventually dies and falls down with a thud. Granted, the first option is a lot more work, but consider the alternative. Every plant should undergo a comprehensive audit, performed by a competent energy specialist which culminates in an established (enforced) conservation program. In lieu of a formal program, here are some basic actions that can be taken.

Daily

- switch off safety/security lights at daybreak
- utilize lighting/heating/air-conditioning only as required
- take advantage of day lighting situations in lieu of electric lighting
- pre-adjust thermostats for comfort cooling and heating
- set back thermostats later in the day prior to exiting
- keep doors/windows closed; make use of venetian blinds

On weekends

- turn off all but safety/security lighting
- set back heating/cooling thermostats
- check settings and accuracy of HVAC controls
- de-energize water heaters and domestic water circulating pumps

Seasonally

- thoroughly test, inspect, and adjust all associated equipment and controls in accordance with manufacturer's recommendations
- turn off heating and in unoccupied spaces
- in hot weather, take advantage of night air to pre-cool the building
- thoroughly winterize buildings and equipment
- adjust timers to accommodate the seasonal temperatures and hours of daylight
- cover or remove window air conditioners at the onset of the heating season

Equipment Operation/Maintenance

You purchase natural gas, oil or coal for your boilers, electricity for your motors and lights, water for processes and consumption, fuel oil for your generators, and gasoline for your vehicles and porta-pumps. Here are some general rules of thumb; followed by a few system specific suggestions on how to manage it all.

- make certain that equipment is properly sized for the connected load
- if it's not being used, turn it off!
- when purchasing new, buy the more energy efficient model
- lower outputs only to needed levels
- avoid demand charges by operating your equipment during "off peak" hours
- lower settings to reduce consumption (temperature, pressure,

speed...)
- run at full capacity; don't overload, but eliminate partial loading
- keep all moving parts clean and well lubricated
- adjust linkages and alignments and replace worn parts
- automate manual controls; calibrate existing controls and devices
- schedule equipment for overhaul and shutdowns; perform internal inspections
- utilize waste heat from processes to preheat fluid or warm spaces
- set controls on refrigeration equipment only as low as necessary
- turn machines off when not in use; don't leave stand-by units energized
- keep nozzles, intakes, drains and orifices unclogged and free flowing
- install heat reclamation devices and eliminate hot water leaks and drips
- reduce water temperatures in hot water systems to acceptable levels
- turn off water heaters or circulating pumps when systems aren't being used
- insulate, insulate, insulate!

Energy Management Systems

Without getting into the subjects of energy audits, life cycle costing and rate of return analyses, it's conceded by most that a well conceived energy management program will provide a good return on investment (ROI), with payback periods of 1 to 3 years as the acceptable norm. But before you go running off to purchase yours, it might pay you to hire an expert to walk you through the process. A good analyst will be able to educate you on what's "state of the art" in the energy conservation field, analyze those parts of your systems you don't understand, estimate potential costs and savings, explain possible modifications and suggest computer hardware and software acquisitions. When you contract with one, make sure it's spelled out what specific services will be provided, what reports will be forthcoming (including audits) and how much your operation can reasonably expect to save, over what period of time. Once the audit is out of the way, an energy management system can be chosen which best matches your established needs. Any EMS worth its chips should be able to:

- turn off lights in unoccupied areas
- maintain partial lighting before and after "public" hours

- schedule lighting operation by hour of day, and time of year
- turn off a water heater when appropriate
- eliminate hot water circulation when an area is unoccupied
- maintain HVAC system start-up and set-back schedules
- eliminate unnecessary HVAC use during unoccupied hours
- monitor and control space temperatures/humidity
- provide temperature and pressure readouts on selected equipment

Bear in mind that when you interrupt, modify or otherwise change your existing systems, you take the chance of affecting the productivity, health and comfort of the buildings machinery and occupants, as well as falling out of compliance with applicable codes and standards. There are devices in your domain, the care of which can be affected by the whimsical ways of Mother Nature. For example, you may be thwarted by snow, wind, rain or ice when attempting to service your building's exhaust fans, roof-mounted air conditioning units or other equipment located outside of your building. The electrical gear in your basement may need to dry or be baked out after a flood during which time the sump pumps run continuously.

HVAC SYSTEMS

The components comprising our HVAC systems are most affected by seasonal maintenance requirements; the hearts of those are our boilers and chillers. As functionally different as these two marvels of technology are in form, they are alike in substance. Essentially, they are the same machine. Both are made of metal and insulated, have moving parts, operating controls and safety devices, and both require myriad auxiliaries with which to operate. More significantly, both are heat exchangers. Their difference lies in the fact that one provides heat and the other removes it. The following PM checklist proves that kinship. Notice that the operating inspections, as well as the "cold iron" inspections are identical for each device, but a season apart.

Spring Boiler Maintenance—Fall Chiller Maintenance

- check system piping for leaks
- test all operating controls
- shut down and drain units

- remove and inspect operating controls
- thoroughly clean unit interiors
- inspect heat transfer surfaces
- lubricate all moving parts
- repack auxiliary valves and pumps
- lay-up per manufacturer's guidelines

Spring Chiller Maintenance—Fall Boiler Maintenance
- renew gaskets and observation glasses
- install new filters as required
- tighten loose connections and fasteners
- calibrate gages and operating controls
- replenish fluids per manufacturer's specifications
- inspect system piping for leaks
- lubricate all moving parts
- check operation of auxiliary equipment
- test safety interlocks

Energy Management Improvements

Less than 30 years long ago, cranking out creature comfort was the primary function of commercial HVAC systems. More recently, new standards and regulations, prompted by technological advances and social conscience, have forced us to rethink how, and what we provide as indoor environment. In light of this, not only have we had to improve the maintenance and repair of our systems, we've also been forced to focus strongly on all the ancillary items that bear upon their operations. Lighting alone reflects up to 30% of a commercial building's total energy load. At 3.41 Btu/watt (sensible heat) it's easy to envision how efficiency improvements in this area would ease the burden on your HVAC systems as well as the bottom line. The bulk of the workload on our chillers isn't from lowering temperature but rather from wringing moisture from the air. Extraction of the latent heat loads via desiccant cooling units engineered into the air stream can have the same effect as reductions in the lighting loads. The nice thing about reworking your HVAC systems is that you can venture into an endless number of areas and find opportunities for improving the quality of your building's air, the efficiency of your equipment and management of your energy resources. Whether you are saddled with pneumatic controls or blessed with computerized DDCs (direct digital controls), and regardless if you will do the work yourself or hire it out, great strides can be made to get your house in order.

Typical Opportunities for Improvements

HVAC Systems

- Install variable speed pumps, fans and drives.
- Optimize chiller systems by reducing condenser water temps.
- Take advantage of free cooling opportunities.
- Install and calibrate accurate sensors and controls.
- Install heat recovery auxiliaries.
- Replace inefficient units with high efficiency models.

Energy Management Systems

- Install an automated energy management system.
- Convert to dual fuel and install oxygen trim systems.
- Establish a good set of pressure and temperature parameters.
- Eliminate incandescent lamps and install energy efficient ballasts.
- Monitor valve positions, motor amperages and pressure drags.
- Install lighting reflectors and occupancy sensors.

Other—Miscellaneous

- Perform a comprehensive energy analysis.
- Establish a steam trap testing and kit repair program.
- Frequently clean air handlers and ductwork.
- Consider installing cogeneration and thermal energy storage capabilities.
- Utilize low pressure drop filtration.
- Insulate and lubricate.

EQUIPMENT DIAGNOSTICS

Electrical devices, rotating equipment, fired and unfired pressure vessels, storage tanks and piping systems represent capital investments in any facility operation. Their failure can result in serious or even catastrophic consequences. It is imperative that their design and construction meet the needs of the physical plant where they are installed, and it is important that they be protected from the damage mechanisms they'll be subjected to throughout the course of their service life. The organizational impact of proper equipment specifications and relevant operational diagnosis result in significant and measurable improvements in plant systems processes and operations, including superior levels of systems integrity, safety, integration, profitability, reliability, availability and maintainability. See Figure IV-1.

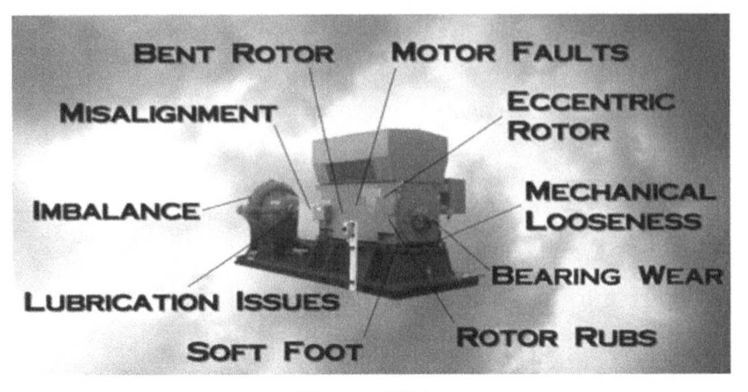

Figure IV-1
Common Equipment Maladies

Reliability Focus

Mechanics and engineers responsible for physical plant operations and maintenance are increasingly incorporating a reliability focus into their strategic and tactical plans and initiatives. This trend is affecting numerous functional areas, including machine/system design and procurement, plant operations and plant maintenance. A reliability engineering focus is increasingly being employed to assure the plants and equipment. The most relevant and practical of these methods include reliability calculations for failure rate, mean time between failures (MTBF), availability, identifying failure time-dependencies and developing an effective field data collection system. Mechanics use different tools for determining the source of a machine's malfunction. Just tearing apart an engine, pump or generator looking for problems is not the most efficient repair process. Listening, measuring, observing and testing helps pinpoint the problem and then dictates the best course of action. Sound diagnostics can reveal that what appears to be a big problem, can be easily fixed replacing a small part. In other situations, what is seemingly a small concern can be a symptom of a much larger problem requiring a major overhaul. Although machinery diagnostics technologies have grown sharply over the last couple of decades, most are cost prohibitive and not readily available to the average maintenance technician. Mechanics still use what could be considered old-fashioned tools for mechanical diagnostics, such as a stethoscope to hear the interior of an engine while it idles; strobe lights to indicate if rotating shafts, gears and other parts are at the correct revolutions per minute by viewing timing marks (one rotation with every strobe blink and the part appears motionless—if the strobe light and part are not synchronized, the part appears to move); and multi-meters for diagnosing electrical circuits. First and foremost, human senses are the most vital and primary diagnostic tools.

Incident Investigations

The purpose of problem solving, troubleshooting and root cause analysis is to determine why a particular issue occurred and to identify specific actions to be taken to prevent negative consequences from re-occurring. A complete investigation program defines the overall approach for a single person, group, site or even an entire company. Five of the basic elements are methodology, measurement, documentation, process and review. See Figure IV-2.

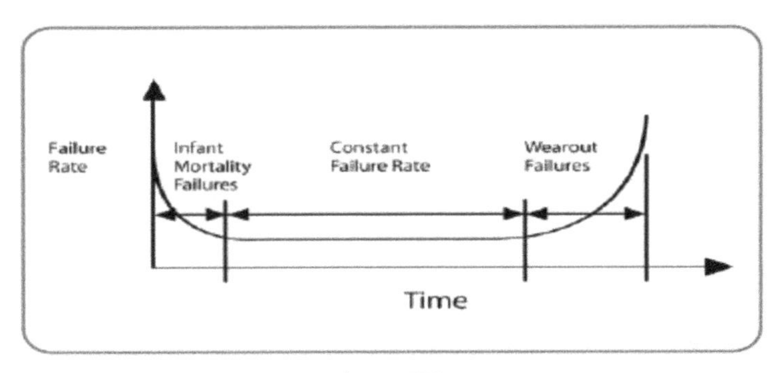

Figure IV-2
Failure Rates over Time

The Mechanic's Focus

Mechanics diagnose system and equipment problems based on symptoms. They know, for example, that if a part is worn, it will be thinner or lose weight as material is lost to friction or some other mechanical process. An experienced mechanic can feel vibrations, hear unexpected noises and observe if different moving parts are functioning correctly. The best diagnostic tool is running the machine and watching, feeling and listening. This simple process can reveal a host of problems to a mechanic trained to interpret the information. Mechanics measure processes such as engine exhaust, fuel levels and quality, coolant levels and temperatures, lubrication flow and power output. Well-equipped technicians use sensing tools and gauges attached to different test fittings to allow access to the system and electronic or analog gauges to reveal how well the various processes perform. Mechanics always measure different working parts and devices and how they perform in accordance with manufacturer's specifications.

Diagnostic Tools

Technicians use micrometers, calipers and scales to accurately measure widths and weights. Sensing tools and gauges attached to different test fittings allow access to the system, and electronic or analog gauges reveal how well the various processes perform. Infrared cameras show heat variations on surfaces to detect where heat is being lost or gained. Diagnostic tools run the gamut of small, unsophisticated, inexpensive and easy to use hand-held devices, to large (cumbersome), expensive, overly sensitive and complicated digital sensors. Following are some of the more traditional diagnostic instruments commonly found in a mechanic's/technician's toolbox.

Hydrometer

A hydrometer is an inexpensive float-type device used for determining the density of a liquid, such as to measure the concentration of sulfuric acid (specific gravity) of battery electrolyte (battery acid). From this reading you can easily and accurately determine a non-sealed battery's state of charge. It has a glass barrel or plastic container with a rubber nozzle or hose on one end and a soft rubber bulb on the other. Inside the barrel or container, there is a float and calibrated graduations used for the specific gravity measurement. See Figure IV-3.

Hygrometer

A hygrometer is an instrument used to measure the moisture content or the humidity of air or any gas. The best known type of hygrometer is the dry- and wet-bulb psychrometer, best described as two mercury thermometers, one with a wetted base, one with a dry base. The water from the wet base evaporates and absorbs heat causing the thermometer reading to drop. Using a calculation table, the reading

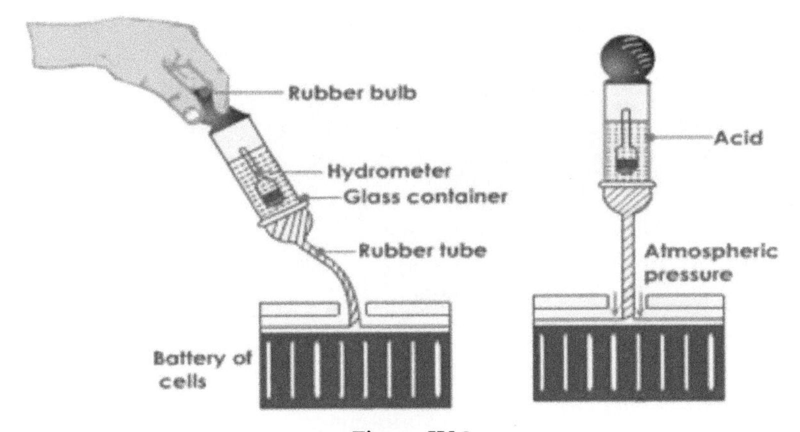

Figure IV-3
Float-type Hydrometer

from the dry thermometer and the reading drop from the wet thermometer are used to determine the relative humidity. See Figure IV-4.

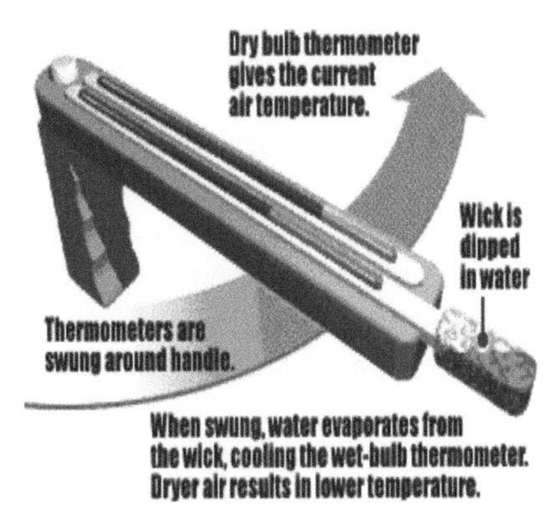

Figure IV-4
Sling Psychrometer

Digital Multi-meter

Multi-meters are digital electronic devices used to take electrical measurements in AC (alternating current) and DC (direct current) circuits. They combine the functions of a voltmeter, ammeter, and ohmmeter to measure continuity, resistance, voltage, current, capacitance and temperature. See Figure IV-5.

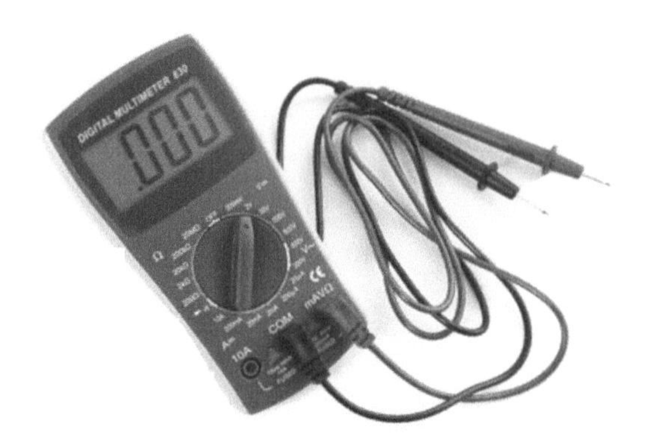

Figure IV-5
Digital Multi-meter

Note: The most common mistake when using a multi-meter is not switching the test leads when switching between current-sensing and any other type of sensing (voltage, resistance). It is critical that the test leads be in the proper jacks for the measurement you are making.

Thermometers

Thermometers come in a wide variety of types (glass with mercury or alcohol/direct reading/digital), temperature scales (Celsius, Kelvin, Fahrenheit and centigrade), ranges (absolute zero to thousands of degrees) and prices ranging from less than $10 to hundreds. What's important is that you purchase one that is reliable, relevant to the work to be performed, able to be calibrated and in your price range. See Figure IV-6.

Stroboscopes

Strobe lights (or stroboscopes) are used to study the behavior of rotating machinery (such as over-speed trips), oscillating vibration of components or structures, and for a lancing. Many times the exact running speed of a machine is unknown. By tuning the strobe and "freezing" motion, the speed will be displayed on the digital readout. The strobe light can be a very useful tool for helping determine the causes of rough running machinery, vibrating brackets or pipe hangers, and for adjusting machinery that is supposed to vibrate as in the case of non-rotating product conveyors. If the strobe light has the ability to measure phase it can

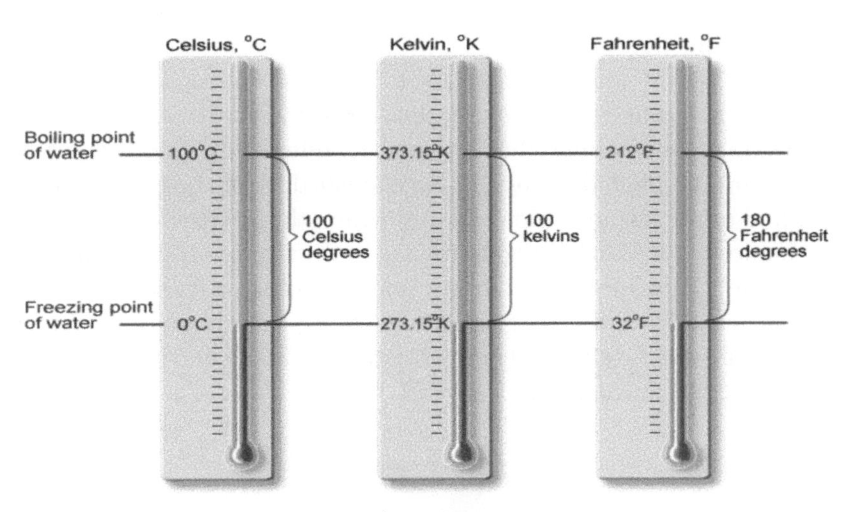

Figure IV-6
Thermometer Temperature Scales

be used for balancing, verifying alignment, checking for looseness and platform or piping system motion studies. See Figure IV-7.

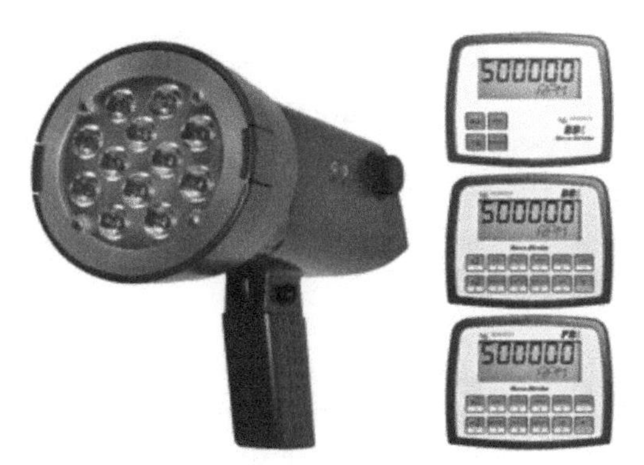

Figure IV-7
Stroboscope

Note: Belts on belt-driven machinery can be examined for irregular operation using a strobe light. For example, in the case of multiple belts, a loose belt will exhibit more motion than belts that have the proper amount of tension. This condition will cause vibration. The strobe light is tuned to the belt speed.

Refrigeration Gauges

Refrigerant gauge manifold sets are used to test the operating condition and cooling status of freezers, box refrigerators, air-conditioners, walk-in coolers, and other systems, operated with various types of refrigerant. Problem diagnosis is accomplished by reading the amount of liquid or refrigerant vapor pressure within the system. Refrigeration mechanics should be thoroughly trained in the safe use and handling of refrigerant (types) and adequately trained in the use of these gauges for testing, leak checking, charging and releasing refrigerants from these cooling devices. Adding and removing refrigerant should only be done by qualified technicians. See Figure IV-8.

Figure IV-8
Refrigerant Gauges

HARMONICS

Equipment with sensitive electronic circuits (digital clocks, VCRs, computers, data terminals) may experience memory loss, system malfunction and even component failure due to electrical power source disturbances. (Sags, surges and harmonics are some common types of disturbances). Harmonics is a term used to explain currents and voltages that have multiplied within an electrical system. Certain orders of harmonics may cause serious equipment and system problems. If the harmonic levels are high, they may cause interference to control and communication lines, heating of AC motors, transformers and conductors, higher reactive power demand (poor power factor), mal-operation of sensitive electronics, overloading of shunt capacitors, and higher power loss. Some utility companies consider imposing penalties on their customers who inject excessive harmonics into the power distribution system (even when their power factor is good). Harmonic distortion disrupts operations, especially productivity and throughput. There are three major classes of harmonic producing devices:

Ferromagnetic (Magnetizing) Device

A coil wound around an iron core, such as in a transformer or motor. These devices normally do not present a problem unless resonant conditions exist where they can amplify the harmonics present in the system.

Electronic Rectifiers and Inverters

Examples are computers, adjustable speed drives, and UPS systems.

Arcing Devices

Examples include fluorescent vapor lighting and arc welders.

Causes

- Adjustable speed drives
- Variable frequency drives
- Electronic welding equipment
- Transformers and generators
- UPS and storage systems
- Medical imaging equipment
- Lighting controls/dimmers
- Computers, copiers and scanners

Effects

- Interference with telephones and communications systems
- Overheated conductors, bus bars, and switch-gear
- Tripped or arcing circuit breakers
- Inaccurate readings from meters and instruments
- Overheated motors
- Breakdown of insulation
- Reversed torque on AC motors
- Reduced equipment life

Typical Harmonic Problems and Solutions

Table IV-1
Harmonics Solutions Table

Symptom	*Cause*	*Solution*
High voltage distortion, no harmonic source near equipment	Capacitor bank in a resonance condition, harmonic currents drawn to bank	Locate source of harmonics, relocate capacitor bank, change capacitor bank size, convert/add filter
High voltage distortion, exceeds limits frequencies	Network is in a resonance condition with one or more dominant harmonic settings (kvar steps), add filters	Locate source of harmonics, move capacitor bank, change controller
Distortion is intermittent, comes and goes at similar intervals	Harmonics generated from a planned load (operation of process), industrial environments	Locate the source, install filters
Capacitor blown fuses, capacitor failure, high harmonics present	High frequency resonance, with high currents (fuses), peak voltage due to a 3rd or 5th order resonance condition (capacitors)	De-tune the network, change capacitor size
Power transformer overheating below rated load, and machinery	Excessive harmonic currents (transformer), high voltage distortion	De-tune or change the capacitor equipment size (transformer), determine the source, install filters if necessary

Measurement

Assembling and implementing a program for improving harmonics in electrical systems can be labor intensive and expensive. The use of harmonic measurement instruments and analyses of harmonic measurement require a high level of sophistication. It is recommended that outside resources and manpower be brought in for this type of work. The purpose of harmonic measurements is to monitor existing values of harmonics and check against recommended or admissible levels, test equipment which generates harmonics, diagnose and trouble-shoot situations where the equipment performance is un-acceptable to the utility or to the user, observe existing background levels, and track the trends in time of voltage and current harmonics patterns.

> **Note**: If in-house personnel are utilized in the program, it is recommended that they undergo special training before being assigned to making such measurements.

Three generic types of instruments used for harmonic measurements are:

Oscilloscopes

The display of the voltage wave-form on the oscilloscope gives immediate qualitative information on the degree and type of distortion. Sometimes cases of resonances are readily identifiable through the multiple peaks present in the current wave. See Figure IV-9.

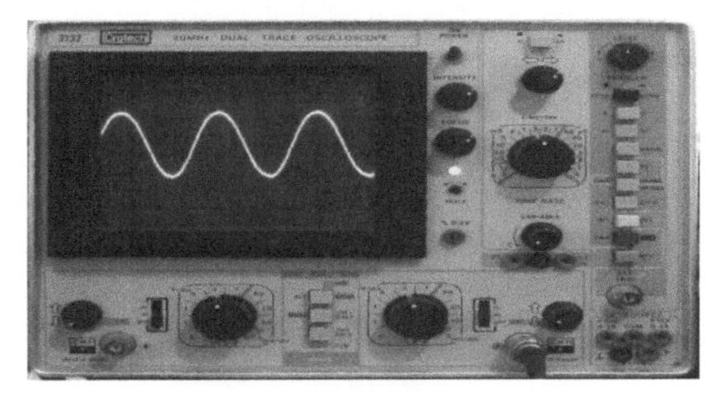

Figure IV-9
Oscilloscope

Spectrum Analyzers

These instruments display the signal as a function of frequency. A certain range of frequencies is scanned, and all the components, harmonics and inter-harmonics of the analyzed signal are displayed. The display format maybe a CRT or a chart recorder. For harmonic measurements, the harmonic frequencies must be identified by reference to the fundamental frequency. A wide range of analog and digital types of spectrum analyzers are available on the market. See Figure IV-10.

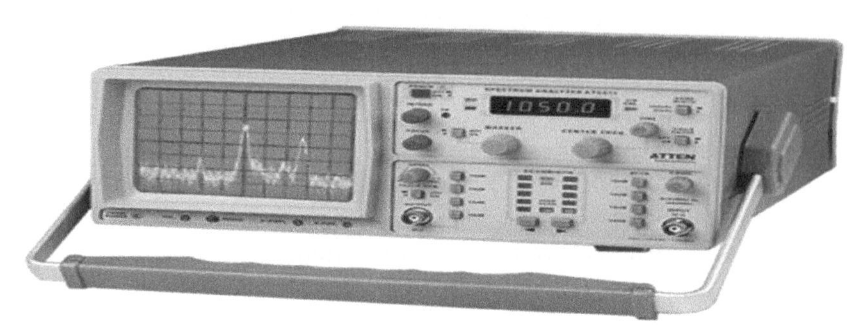

Figure IV-10
Spectrum Analyzer

Wave Analyzers

Harmonic analyzers or wave analyzers measure the amplitude (and also phase angle in more complex units) of a periodic function. These instruments provide the line spectrum of an observed signal. The output can be recorded or can be monitored with analogue or digital meters. See Figure IV-11.

Figure IV-11
Wave Analyzer

Solutions

Many standard harmonic reduction solutions are available, including reactors, isolation transformers, filters and active devices.

All have strengths and weaknesses and should be considered carefully in the context of your particular harmonics problems. Harmonic mitigation is especially important when power factor correction capacitors are already installed in your facility, or if you plan on adding them in the future. Even though capacitors do not create harmonics, they resonate and amplify existing harmonics. Adding passive harmonic filters to your capacitors will protect your capacitors from being damaged by existing harmonics. There are two approaches to harmonic mitigation:

Attending to the Symptoms

In some facilities, it's best (and easiest) to treat the symptoms of harmonics. If your only problem is neutral conductor overheating, you can increase the conductor's size. If your transformers are overheating, you can install special K-rated transformers designed to better tolerate harmonics. You can relocate harmonic-producing loads around your facility to balance the harmonics and produce a better sine wave.

Correcting the Cause

When treating the source of harmonics, a power quality study or measurements from monitoring equipment normally will show a need for a more complex solution. To reduce the level of harmonics produced by a facility's equipment, impedance may be added by installing line reactors at the source, or passive filters can be installed to eliminate specific harmonic frequencies, or an active filter can be installed to address a broad range of harmonic orders when they are present.

HEAT AND TEMPERATURE

Heat and temperature are not one and the same thing. Heat is thermal energy and temperature is a measure of how hot or cold an object or substance is. There are two kinds of heat, known by their effects (sensible heat and latent heat). The differences can be seen in Figure IV-12. As you can see, it doesn't deal with two kinds of heat, but rather with two of heat's effects. The terms sensible heat and latent heat are not special forms of energy, but they do play a special role in HVAC operations, where they describe exchanges of heat under conditions specified in terms of their effect on a material or a thermodynamic system.

Note: Sensible heat is related to changes in temperature of a solid, liquid or gas (as can be measured on a thermometer), without a corresponding change of state.

Note: Latent heat refers to changes in state of a solid, liquid or gas (i.e., ice to water, water to vapor, vapor to water, water to ice) without a corresponding change in temperature.

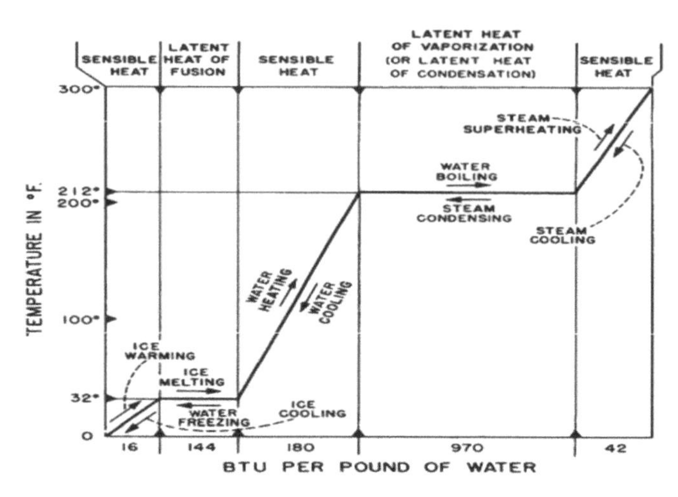

Figure IV-12
Heat Relationships

Specific Heat

Specific heat is the number of Btus of heat required to raise or lower l pound of substance by 1°F. Water, being so abundant, is used as the basis for measurement having a specific heat of 1.0. Cast iron's is 0.1298, and ice has a specific heat value of 0.504. **Note**: It takes one Btu of heat to raise the temperature of 1 pound of water by 1°F (at sea level). The relationship of sensible and latent heat can be seen by studying Figure IV-12. It depicts the amount of heat required in Btus (British thermal units) to both change the state of and raise and lower the temperature of one pound of water, between states, and from 0°F to 300°F. When the flow of heat is reflected in a temperature change, we say that sensible heat has been added to or removed from the substance (heat that can be sensed or felt). When the flow of heat is not reflected in a temperature change, but is reflected in the changing physical state of a substance, we say that latent heat (also referred to as hidden heat) has been added or removed.

Sensible Heat

Sensible heat is potential energy in the form of thermal energy or heat. The thermal energy can be transported via conduction, convection, radiation or by a combination thereof. The quantity or magnitude of sensible heat is the product of the body's mass, its specific heat capacity and its temperature above a reference temperature. In many cases the reference temperature is inferred from common knowledge, i.e., room temperature. When the flow of heat into or from a substance occurs without benefit of a change of its physical state, we say that sensible heat has been added to or removed from the substance. We can substantiate the transfer through the use of thermometers or other temperature-sensitive devices which indicate corresponding increases or decreases in the temperature of the substance on their scales.

Latent Heat

Latent heat is the amount of energy in the form of heat released or absorbed by a substance during a change of phase (i.e., solid, liquid, or gas). Two latent heats (or enthalpies) are typically described: the latent heat of fusion (melting), and the latent heat of vaporization (boiling). The names describe the direction of heat flow from one phase to the next: solid → liquid → gas. Sometimes referred to as hidden heat, latent heat is evident only when the physical state of a substance is altered, as when water freezes or ice thaws. Tremendous amounts of heat must be transferred to accomplish changes of state, but your temperature-monitoring equipment will do you little good here, as no rise or fall in temperature takes place during the process. And though the phenomenon is difficult to comprehend, it serves as the basic operating premise for both the steam generating and refrigeration fields. As you can surmise from Figure IV-12, temperature rises and falls when heat is added to or removed from the sensible-heat labeled areas, and large amounts of heat addition and removal are required to facilitate a change of state in the water, with no change in temperature in the latent-heat labeled areas. Using 1 pound of water at atmospheric pressure, Figure IV-12 illustrates that it takes the addition of 1,354 Btus (British thermal units) to raise the temperature of ice at 0°F to steam at 300°F, 240 degrees of which represent sensible heat and 1,114 of which represent latent heat. If the process were reversed, the same amount of heat would need to be removed to cool 300-degree steam to ice at 0°F.

Both sensible and latent heat are observed in many processes while transporting energy in nature. Latent heat is associated with changes of state, measured at constant temperature, especially the phase changes of atmospheric water vapor, mostly vaporization and condensation, whereas sensible heat directly affects the temperature of the atmosphere. Factors that influence sensible heating and cooling loads include glass windows or doors, sunlight striking windows, skylights, or glass doors (heating the space), exterior walls, partitions (that separate spaces of different temperatures), ceilings under attic roofs, floors over open crawl spaces, air infiltration through cracks in the building, doors and windows, people in the building, equipment and appliance operations and lights. Other sensible heat gains are taken care of by the HVAC equipment before the air reaches the rooms (system gains). Two items that require additional sensible cooling capacity from the HVAC equipment are ductwork located in an unconditioned space and ventilation air (air that is mechanically introduced into the building).

Temperature Scales

All substances (even the coldest) have some heat within them. Cold is nothing more than a relative term invented for the masses to help them perceive the thermal energy contained in substances through the human sense of touch. Heat is manifested within a substance by the vibration of its molecules. The faster the molecules vibrate, the more heat a substance contains; conversely, the slower the molecules vibrate, the less heat it contains. As we've already learned in the section on heat, a change in the physical state of a substance is a visual indication that latent heat transfer has taken place within a body. Temperature-sensitive devices tell us when sensible heat has been transferred, and their temperature scales literally indicate to what degree. See Figure IV-13.

The two most commonly used temperature scales are the centigrade and Fahrenheit scales. Certain points on the scales are important to us in plant engineering (specifically boiling and freezing temperatures). The Fahrenheit scale is a thermometric scale on which the boiling and freezing points of water register at 212°F and 32°F above its zero

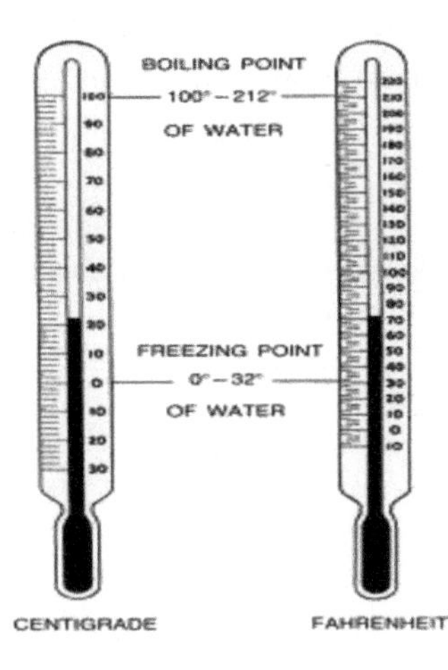

Figure IV-13
Commonly Used Temperature Scales

mark, respectively, under standard atmospheric pressure at sea level. The centigrade (or Celsius scale) is a thermometric scale on which the boiling and freezing points of water register at 100°C and 0°C respectively, under standard atmospheric pressure at sea level. To illustrate these points, in Figure IV-13, lines have been drawn across the scales where they register the boiling and freezing temperatures of water and absolute zero for comparison. Other scales are available as well.

The Rankin Scale is an absolute-temperature scale, the unit of measurement of which equals a Fahrenheit degree. It registers the freezing and boiling points of water at 492°F and 672°F respectively. And the Kelvin Scale is an absolute temperature scale, the unit of measurement of which equals a Celsius degree. It registers the freezing and boiling points of water at 273°C and 373°C respectively. The latter scales are not normally found in the physical plant as they are mostly used for tracking more extreme temperatures. Although science has never accomplished the complete removal of all the heat from a substance, it is a generally accepted theory that absolute zero is a hypothetical temperature value assigned to a condition characterized by the complete absence of heat in a substance when, it is thought by some, all molecular activity ceases. Absolute zero on the Fahrenheit scale registers at approximately –460°F and on the Celsius scale at approximately –273°C.

Temperature Conversion

Work in the physical plant sometimes calls for the need to convert degrees on the Fahrenheit scale to their centigrade equivalent (or vice versa). To do so, use the formula, C = 5/9 x (F – 32). To convert degrees on the centigrade scale to their equivalent Fahrenheit temperatures, use the formula, F = 9/5 C + 32.

Example:

Problem: What is the temperature equivalent of –40°C on the Fahrenheit scale?

Solution: 9/5 x C (deg. Celsius) + 32 = (9/5 x – 40) + 32 = (– 72) + 32 =

Answer: –40°F

INDOOR AIR QUALITY (IAQ)

The epitome of good indoor air quality includes the introduction and distribution of adequate ventilation air, control of airborne contaminants, maintenance of acceptable temperature and maintenance of relative humidity. The U.S. EPA ranks indoor air pollution among the top five environmental risks to public health. Unhealthy indoor air is found in up to 30 percent of new and renovated buildings. Sick building syndrome (SBS) and building related illness (BRI) have become more common in the workplace, increasing building owner and employer costs due to sickness, absenteeism, and increased liability claims. Good air quality is an important component of a healthy indoor environment. It enhances occupant health, comfort and workplace productivity. A healthy indoor environment is one in which the surroundings contribute to productivity, comfort, and a sense of health and well-being. The indoor air must be free from significant levels of odors, dust and contaminants and circulate to prevent stuffiness without creating drafts.

Equipment Related Problems

Indoor air quality (IAQ) problems can result from many causes. Failure to respond promptly and effectively to IAQ problems can have consequences that result in health problems, reduced productivity and life-threatening conditions. Some (but not all) contributing causes include inadequate humidity control, inadequate air distribution to drafty and stuffy rooms, dry sewage drain traps allowing sewer gas escape, entry of vehicle exhaust into building openings, inadequate temperature control, and high concentrations of operating equipment.

Classifications

Poor indoor air quality can cause human illness, which in turn may result in increased liability and expense for building owners and operators. It can also lead to lost productivity of building occupants, resulting in economic losses. Health problems that can result from poor indoor air quality may be short-term to long-term, and range from minor irritations to life-threatening illnesses. They are classified as follows:

Sick-building Syndrome (SBS)

SBS describes a collection of symptoms experienced by building occupants that are generally short-term and may disappear after the individuals leave the building. The most common symptoms are sore throat, fatigue, lethargy, dizziness, lack of concentration, respiratory irritation, headaches, eye irritation, sinus congestion, dryness of the skin (face or hands), and other cold-, influenza- and allergy-type symptoms.

Building-related Illnesses (BRI)

BRIs are more serious than SBS conditions and are clinically verifiable diseases that can be attributed to a specific source or pollutant within a building. Examples include cancer and Legionnaires' disease (from cooling tower operations).

Multiple Chemical Sensitivities (MCS)

More research is needed to fully understand these complex illnesses. The initial symptoms of MCS are generally acquired during an identifiable exposure to specific VOCs. While these symptoms may be observed to affect more than one body organ system, they can recur and disappear in response to exposure to the stimuli (VOCs). Exposure to low levels of chemicals of diverse structural classes can produce symptoms.

Contributing Factors

Air quality in poorly maintained buildings can deteriorate quickly. Materials, products, furniture, and HVAC systems need regular maintenance, cleaning, and inspections to ensure that they function as designed and to prevent indoor contaminants from developing in those locations. Other potential problems result from the use of pesticides, microbial growth caused by moisture within the building, and the emission of sewer gas where floor drains are concealed.

Building Envelope

The envelope controls the infiltration of outside air and moisture, and may include operable or inoperable windows.

Ventilation Systems

Acoustical materials in heating, ventilating, and air-conditioning (HVAC) systems may contribute to indoor air pollution.

Maintenance

Lack of maintenance allows dirt, dust, mold, odors and particles to increase.

Electric and Magnetic Fields (EMF)

The possible health effects of electric and magnetic fields generated by power lines and electric appliances.

Strategies

Develop and provide the building operators with complete operations and maintenance manuals and a plan for appropriate system operation training. After the tenants have occupied a new or remodeled building, implement post-occupancy building commissioning and flush out the building as necessary to fine-tune the building systems under normal operating conditions. Develop a plan to provide post-occupancy building commissioning on a regular basis every few years. Strategies for building maintenance include building commissioning and design documentation that describes: the building and its systems; the function and occupancy of each individual room or space; normal operating hours, and any known contaminants and hazards; schematic drawings of the building systems (indicating equipment types), their locations, and their maintenance and inspection points; locations of system manuals and commissioning reports; as-built drawings, water treatment logs, inspection reports, and training manuals; performance criteria and operating setpoints for each operating unit (including domestic water system and normal humidity levels); sequence of operations for equipment and systems, along with seasonal startup and shutdown procedures; daily operating schedules of all systems; preventive-maintenance and inspection schedules for equipment; test-and-balance report and airflow rates (listed by area); required outdoor-air rates and building pressurization requirements; building IAQ inspection checklists; equipment maintenance checklists; procedure for documenting and responding to occupant complaints; and IAQ documentation.

Management

IAQ investigations are conducted by industrial hygienists or other qualified personnel, following guidelines published by the American Conference of Industrial Hygienists (ACGIH), the American International Health Alliance (AIHA), the American National Standards Institute (ANSI), the American Society of Heating, Refrigeration, and Air Conditioning Engineers (ASHRAE), USEPA, OSHA, NIOSH; or other federal, DOD, state and local municipal organizations. A typical IAQ investigation consists of occupant interviews, on-site physical inspection, and non-destructive testing. These steps help in determine whether the problem is biological, non-biological, building-related or not, transient or chronic, etc., and establish appropriate remediation strategies such as educating maintenance mechanics on measures they can take to help maintain acceptable IAQ in their work areas, not making unauthorized modifications to the heating, ventilation, and air conditioning (HVAC) systems (i.e., blocking off vents, removing ceiling tiles). Some IAQ problem situations require immediate action. Other problems are less urgent, but all merit a response. Carbon monoxide poisoning (for example) is a potentially life-threatening illness, requiring sources of combustion gases to be investigated. Proper cleaning and disinfection procedures must be used to prevent the growth of mold and bacteria that could cause serious indoor air quality problems. Inadequately maintained humidifiers can promote the growth of biological contaminants. Clean equipment thoroughly, and consider modifying maintenance practices. Even when the symptoms described suggest an IAQ problem that is not life-threatening, it would be wise to respond promptly.

Ventilation and Exhaust Systems

Ventilation systems should be designed to prevent dispersion into the air, or the drawing through the work area, of dusts, fumes, mists, vapors, and gases in concentrations causing harmful exposure. Their engineering controls should reduce or eliminate airborne contaminants created by portable equipment (such as drills, saws, and grinding machines) in concentrations exceeding acceptable safe limits, and the efficiency of the control systems should be periodically verified. Ventilation systems should be designed, installed, operated and maintained in such a manner as to ensure the maintenance of the volume and velocity of exhaust air sufficient to gather contaminants and safely transport them to suitable points for removal.

Duration of Operation

Ventilation systems should be operated continuously during operations when persons are exposed to airborne contaminants or explosive gases at or above acceptable safe limits as defined in or specified by applicable standards and regulations; and they should remain in operation for a time after the work process or equipment has ceased, to ensure the removal of any contaminants in suspension or vaporizing into the air.

Contaminant Disposal

Dusts and refuse materials removed by exhaust systems or other methods should be disposed of in a manner that will not create a hazard to employees or the public and in accordance with federal, state and local requirements.

Biological Growth

Molds and their spores are found in all indoor and outdoor environments. They tend to thrive, however, only when there is an available food source and when placed in a warm and moist environment. Mold growth indoors is usually preceded by a water-intrusion event (e.g., broken pipe, infiltration of rain or melting snow, sewer backup, etc.). Porous building surfaces (carpeting, drywall, ceiling tiles, etc.) and other non-structural material (paper, books, cardboard, etc.) provide a food source in the form of cellulose. Fungi in buildings may cause or exacerbate symptoms of allergies. Correct all water intrusion events. Clean water left for more than 24-48 hours can lead to mold growth. Events involving sewage backflows are very serious. If the problem is strictly intrusion of clean water with no other complaints, the remediation can usually be limited to quick and thorough drying of the impacted area. If sewage or contaminated water was involved, some removal activities (i.e., removal of wet carpets, drywall, etc.) may also be implemented to prevent later problems with biological growth.

Avoiding Problems

You can help to avoid IAQ problems in your work area by preparing a maintenance plan with a schedule and budget for the HVAC systems, building materials, and furniture that involves activities such as maintaining water treatment, proper filter changes, lubricating dampers, removing standing water and moisture, cleaning condensate pans, coils and supply-air ducts, equipping vacuum cleaners with HEPA filters, periodically deep cleaning carpets, regularly cleaning and vacuuming furniture and regularly inspecting for microbial growth. Other action items might include unblocking shut vents or building returns, routinely cleaning work areas and avoiding the use of portable humidifiers.

POWER FACTOR

Power factor is a measure of how efficiently electricity is consumed in an AC (alternating current) electrical power system. Various types of power provide us with electrical energy needed to operate equipment, including:

Working Power

The "true" power used to perform the work in all electrical appliances; it is measured in kW (kilowatts). KW is also called actual power, active power or real power. It is the power that actually powers the equipment and performs useful work.

Reactive Power

Inductive loads require reactive power to generate and sustain a magnetic field, in order to operate; that is measured in kvars (kilovolt-amperes-reactive). Inductive loads (which are sources of reactive power) include devices such as:

- Fluorescent lighting
- Transformers
- Induction motors
- Induction generators

Kvar is reactive power. It is the power that magnetic equipment (transformers, motors and relays) need to produce the magnetizing flux.

Apparent Power

Working power and reactive power make up apparent power; that is measured in kVA (kilovolt-amperes). KVA is apparent power. It is the vectorial summation of kvar and kW. It is the ratio of working power to apparent power, or the formula PF = kW/kVA. It is defined as the ratio of the real power flowing to the load, to the apparent power in the circuit, and is a dimensionless number between 0 and 1. To understand power factor, you must have an understanding of the definition of these basic terms, and the power factor triangle used for determining power factor values. The three circuit elements which make up the electrical power consumed in an AC circuit can be represented by the three sides of a right-angled triangle, known commonly as a power triangle. See Figure IV-14.

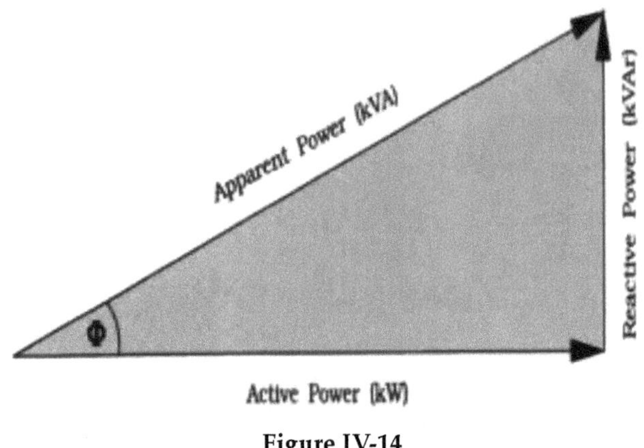

Figure IV-14
Power Factor Triangle

Basics

Without getting into a higher math scenario (phase angles, cosines, sinusoidal waves and such), real power is the capacity of the circuit for performing work in a particular time. Apparent power is the product of the current and voltage of the circuit. Due to energy stored in the load and returned to the source, or due to a non-linear load that distorts the wave shape of the current drawn from the source, the apparent power will be greater than the real power. In an electrical power system, a load with a low power factor draws more current than a load with a high power factor for the same amount of useful power transferred. The higher currents increase the energy lost in the distribution system, and require larger wires and other equipment. Because of the costs of larger equipment and wasted energy, electrical utilities will usually charge a higher cost to industrial or commercial customers where there is a low power factor.

Power factor is best expressed as active power (kW)/apparent power (kVA). The ideal power factor is unity—or one. Anything less than one, (or 100% efficiency), means that extra power is required to achieve the actual task at hand. A low power factor indicates inefficient use of electrical power. Maintaining a higher power factor benefits both the utility and consumers through lower power delivery system losses and greater customer equipment operating efficiency.

Example: A 100 kW motor operating at a power factor of 0.80 has a total apparent power requirement of 125 kVA (100 kW/0.80). If the power factor can be improved to 0.95, the total power draw from the supply will be reduced; i.e., 100 kW/0.95 = 105 kVA.

Power Factor Correction

Power factor correction is the term given to a technology employed to restore power factors to as close to unity as is economically possible. The benefits associated with that action are lower utility bills, reduced operating costs, efficient use of utility power, reduced equipment losses, quick returns on investments (ROI), cost savings over equipment life and reduced power consumption. See Figure IV-15.

> **Note**: An inductive load requires a magnetic field to operate, and creating such a magnetic field causes the current to lag the voltage (that is, the current is not in phase with the voltage). Power factor correction is the process of compensating for the lagging current by applying a leading current in the form of capacitors. That way, power factor is adjusted closer to unity, and energy waste is minimized.

Installing Capacitors

Installing capacitors decreases the magnitude of reactive power (kvar), thus increasing your power factor. Capacitors store kvars and release energy opposing the reactive energy caused by the inductor. The presence of both a capacitor and inductor in the same circuit results in the continuous alternating transfer of energy between the two. Thus, when the circuit is balanced, all the energy released by the inductor is absorbed by the capacitor.

Capacitor Types

Simple, small fixed capacitors can be installed at single-motor locations. Larger fixed assemblies can be installed to work with more

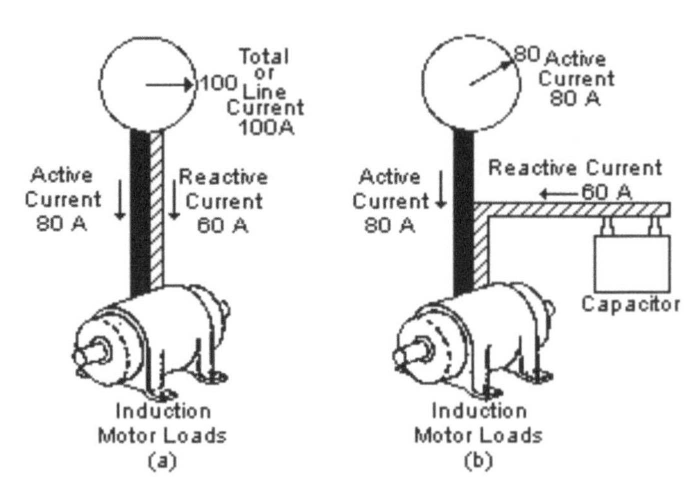

Figure IV-15
Power Factor Correction, Using a Capacitor

than one motor. Still larger automatic switched capacitor systems can be installed for a large sector of a facility, or at the service entrance, to help correct the power factor of an entire facility. Capacitor systems may be integrated with switchgear, retrofitted, or installed in a match-and-line arrangement. Where they are installed in the electrical system has a direct bearing on their cost, benefit and flexibility.

Location	*Cost*	*Benefit*	*Flexibility*
Motor	Low	Fair	Minimal
Feeders	Medium	Good	Better
Entrance	High	Best	Maximum

Note: Low power factor values can be also be improved by minimizing operation of idling or lightly loaded motors, avoiding operation of equipment above its rated voltage and replacing standard motors as they burn out with energy-efficient motors.

Power Factor Calculations

In electrical terms, power factor is the ratio of true power to apparent power expressed as a percentage, which indicates the efficiency of a circuit. In direct current (DC) circuits, the power factor value is always 1.0 (or 100 percent) and their power equation is represented as:

WATTS = VOLTS x AMPERES

In alternating current (AC) circuits, power factor values vary between 0 and 1.0 (or zero and 100 percent) depending on load and circuit characteristics; and their power equation is represented as:

WATTS = VOLTS x AMPERES x POWER FACTOR

Problem:

Determine the power factor of a 1200 kVA diesel-driven generator that develops 1080 kilowatts of power at full load. (K or kilo equals 1000)

Solution:

POWER FACTOR = WATTS/VOLTS x AMPERES

pf (power factor) = kW (kilowatts)/kVA (kilovolt amperes)
= (1080 x 1000)/(1200 x 1000)
= 1080/1200

Answer: 0.9 or 90 percent

PRESSURE AND VACUUM

Maintenance mechanics and operating engineers work with extremes of pressures and vacuum every day, and it appears that not all pressures are created equal. Actually, there aren't different kinds of pressure, just different ways that pressure is represented. Let's review the topic and you'll see what I mean.

Atmospheric Pressure

The weight of all the air that makes up the atmosphere surrounding our planet is brought to bear on its surface as a consequence of the Earth's gravitational pull. Atmospheric pressure is a measurement of the weight of the air over a given area (usually a square inch) taken at some point in a column of air which extends from the Earth's surface to an indefinite point in outer space. This is generally accepted as 14.7 pounds per square inch (psi) at sea level. In other words, each column of air having a 1-inch-square base which extends from the Earth to space contains approximately 15 pounds of air by weight. If you measure a square foot of area, the total weight increases to over 2,000 pounds or approximately a ton of air by weight, since there are 144 square inches in a square foot, and each of them contains a column of air weighing (for the purpose of simplifying the calculation) 15 pounds; i.e., 15 (pounds of air) x 144 (square inches) = 2,160 (pounds of air). That truth established, it then stands to reason that the weight (pressure) of the air would be less at altitudes above sea level and more on areas located below sea level. This accounts for the facts that the air is "thinner" in the mountains and that water boils at temperatures increasingly lower than 212°F as you ascend them. The device used for measuring atmospheric pressure is the barometer. See Figure IV-16.

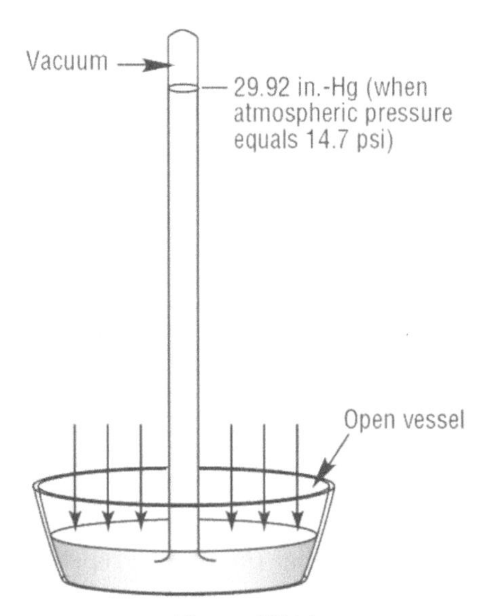

Figure IV-16
Torricelli Barometer

Vacuum

When the pressure existing in a vessel is below atmospheric, the vessel is said to be in a vacuum. Actually, pressure still exists within the vessel, but at a reduced level. Vacuum gages or the vacuum side of compound gages are used to measure this condition. Just as pressure gages measure pressures at or above atmospheric in psi, vacuum gages measure pressures below atmospheric in inches of vacuum. Two inches of vacuum are approximately equivalent to 1 psi. So, for instance, if you should read a vacuum gage and it shows a vacuum of 16 inches, that would indicate that the pressure you were measuring was 8 psi less than the surrounding atmosphere or approximately 7 psi at sea level.

Example

16 (gage reading in inches)/2 (inches per psi) = 8 psi (pressure below atmospheric)

15 psi (atmospheric pressure) minus 8 psi (pressure below atmospheric) = 7 psi (pressure remaining in vessel).

Or Put Another Way

30 (perfect vacuum) minus 16 (gage reading in inches)/2 = 7 psi (pressure remaining in vessel).

As you can see, even though the pressure remaining in the vessel is well below that of the surrounding atmosphere, the vessel still contains pressure. Externally it is being subjected to 15 pounds of pressure on every square inch of its surface, while internally the vessel is being pressurized to the tune of 7 pounds on every square inch of its surface. The vacuum scale reads from zero (atmospheric pressure) to 30 (perfect vacuum). See Figure IV-17.

Figure IV-17
Compound Pressure—Vacuum Gage

Note: A compound gauge is a device that can display both positive and negative (vacuum) pressures. Pressure gauges use pounds per square inch (psi) as the unit of measure. Vacuum gauges, on the other hand, measure force in units of inches of mercury (in Hg). Both of these measurements will be displayed on the face of a compound gauge. The needle in a compound gauge will move clockwise when measuring positive pressure, and counterclockwise when measuring negative pressure.

Gage Pressure

Given a pressure gage which is graduated from zero and read in pounds per square inch (psi), gage pressure is the pressure which is indicated by the gage dial at or above atmospheric pressure. A dial pointing to zero indicates that the pressure being measured is exactly the same as the pressure of the surrounding atmosphere. A dial fixed on 100 psi indicates that the pressure being measured is 100 psi in excess of atmospheric pressure. A dial pointing below zero indicates that the gage is broken or that the pressure being measured is below atmospheric (in the vacuum range), that a problem may exist in your pressure vessel or that an inappropriate gage is being used to measure the pressure within it. See Figure IV-18.

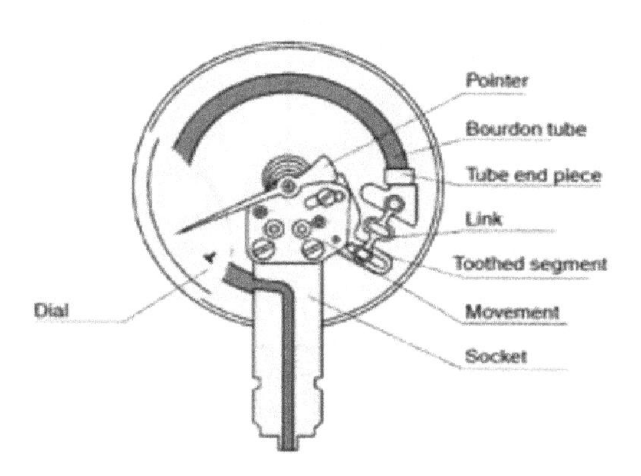

Figure IV-18
Bourdon Tube Pressure Gage

Absolute Pressure

Encompassing the whole range of pressure is the absolute scale. For purposes of mathematical calculation it can be expressed as atmospheric pressure plus gage pressure. Its scale starts at zero (perfect vacuum) and extends forever to infinity. See Figure IV-19.

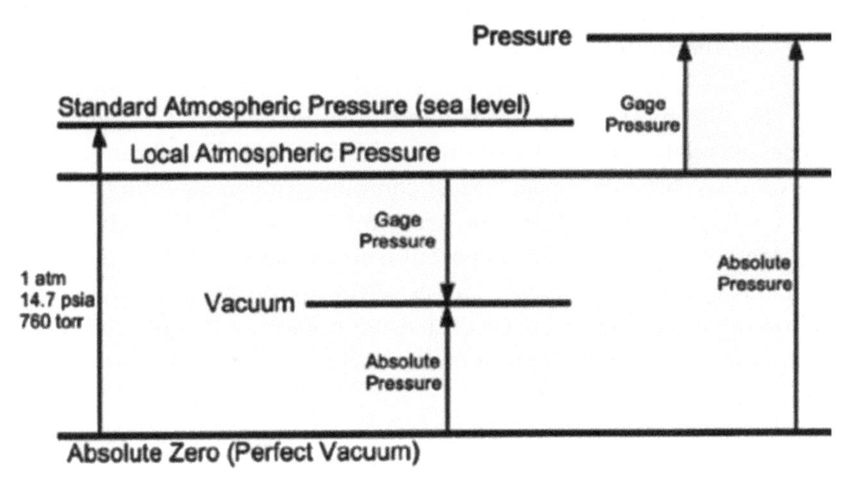

Figure IV-19
Pressure-vacuum Relationship

PSYCHROMETRIC CHARTS

Psychrometry is the measurement of the thermodynamic properties in moist air. As a problem-solving tool, psychrometrics clearly show how changes in heating, cooling, humidification, and dehumidification can affect the properties of moist air. The psychrometric chart is a graphical depiction of the thermodynamic properties which impact moist air. The charts focus on the range of temperatures most suitable for comfort-cooling, heating, ventilation and air-conditioning (HVAC) applications, and are essential to understanding the climate of a facility. Psychrometric charts contain a great deal of information in a relatively compact format. Because of this, many are intimidated when first introduced to it. Once a few basic concepts are understood, the chart is really quite simple to use. "P" charts include all of the following properties for moist air:

- dry bulb temperature
- wet bulb temperature
- relative humidity

- dew point temperature
- humidity ratio
- total heat (enthalpy)
- specific volume

Psychrometric data are needed to solve various problems and processes relating to air distribution. This is accomplished through the use of a psychrometric chart. Most complex problems relating to heating, cooling and humidification are combinations of relatively simple problems. The psychrometric chart illustrates these processes in graphic form, clearly showing how changes affect the properties of moist air. Following are the characteristics that are tracked and a skeletal representation of where they appear on the chart.

Psychrometry Basics

> **Note**: If any two of the seven properties are known, the remaining properties can be obtained

A psychrometric chart may be a troublesome tool for maintenance mechanics to learn to use, but the chart's importance in the analysis of industrial systems involving air and water vapor is undeniable, and reading the chart is an important skill for them to master. The following reflects the seven points that can be read on a psychrometric chart, and the information they provide. See Figure IV-20.

Dry Bulb Temperatures (DB)

Dry bulb temperature is the temperature of a substance as read by a common thermometer. It is an indication of the sensible heat content of the substance. Dry bulb temperatures are shown as vertical lines originating from the horizontal axis on the bottom of the chart.

Wet Bulb Temperature (WB)

The wet bulb temperature is used to measure the water content of moist air. It is obtained by passing air over a thermometer that has a wet wick over its sensing bulb. The drier the air, the more water will evaporate from the wick which lowers the reading on the thermometer. If the air is saturated (100% relative humidity), no water will evaporate from the wick, and the wet bulb temperature will equal the dry bulb tempera-

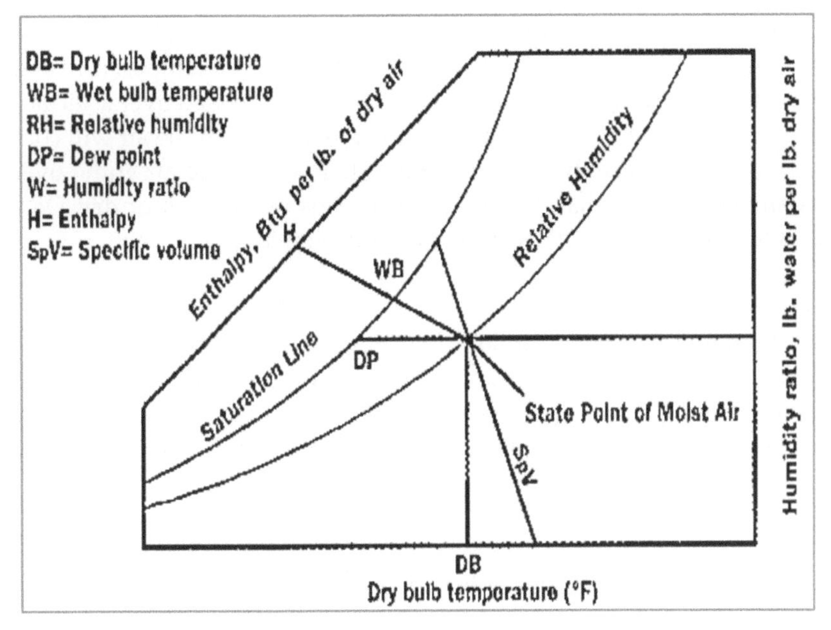

Figure IV-20
Psychrometric Value Lines

ture. Wet bulb lines originate from where the dry bulb lines intersect the saturation line and slope downward and to the right. Wet bulb lines are nearly but not exactly parallel to enthalpy lines.

Relative Humidity (RH)

The ratio of the amount of water vapor in a given sample of air to the maximum amount of water vapor the same air can hold. 100% RH indicates saturated air (the air cannot hold any more water vapor), and 0% RH indicates perfectly dry air. (**Note**: The above definition is accurate for all practical purposes. The correct definition of relative humidity is the ratio of actual water vapor pressure in a sample of air, to the water vapor pressure in saturated air at the same temperature.) The 100% RH line is the saturation line and lines of lesser RH fall below and to the right of this line.

Dew Point Temperature (DP)

The temperature to which air must be cooled before condensation of its moisture will begin. As a sample of air is cooled, its RH climbs until it reaches 100% RH (saturated air). This is the dew point temperature. At saturation, dew point temperature equals wet bulb temperature and dry bulb temperature, and the RH is 100%. If air is passed over a surface that is below the dew point temperature, moisture from the air will condense on the surface. It is the dew point temperature of air going over a cooling coil's fins that determines if the fins will be wet or dry. Dew point temperatures are shown on the saturation line.

Humidity Ratio (W)

Sometimes referred to as "specific humidity," this is the actual weight of water vapor in a pound of dry air. Humidity ratio is expressed in pounds (or grains) of water vapor per pound of dry air. Humidity ratio lines are horizontal on the chart and originate from the vertical axis on the right-hand side.

Enthalpy (H)

This term is used to describe the total heat of a substance and is expressed in Btu per pound. For moist air, enthalpy indicates the total heat in the air and water vapor mixture and is shown as Btu per pound of dry air. Dry air at 0°F has been assigned an enthalpy of 0 Btu/lb. Enthalpy values are found on a scale above and to the left of the saturation line. Lines of constant enthalpy slope downward and to the right and nearly parallel to the wet bulb lines.

Specific Volume (SpV)

The reciprocal of density, specific volume is expressed as cubic feet of air-to-water vapor mixture per pound of dry air. Lines of specific volume start on the horizontal axis and slope upwards and to the left.

Example

Here is an example exercise in reading a psychrometric chart.

Given sling psychrometer readings of DB = 95°F, WB = 76°F, the other five psychrometric properties of this air can be found as referenced in Figure IV-21, where the given condition is plotted as state point A. Following the RH line at that point, it can be interpolated that the relative humidity corresponding to those two values is 42%. Following a horizontal line right to the specific-humidity scale gives W = 104.5 gr of water vapor per pound of dry air. Following the same line left to the saturation line gives DP = 68.6°F. Interpolating between the SpV lines of

14.0 and 14.5 gives an approximate SpV of 14.3 ft^3/lb air. The enthalpy is found by following the WB line left and up to the saturation line and beyond to the enthalpy scale. For the condition given, H = 39.55 Btu/lb of dry air.

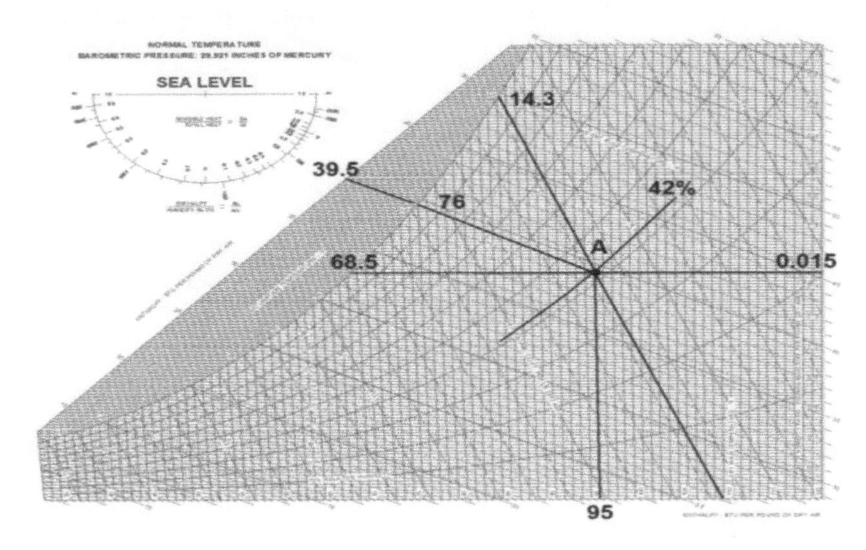

Figure IV-21
Example Chart

PUMP CURVES

High maintenance costs and innumerable operating problems are directly associated with incorrectly sized pumps and accoutrements installed within pumping systems. Too often, maintenance mechanics and operating engineers think of pumps like light bulbs; meaning, you plug them in, and they work. This often leads to operation and maintenance problems that can otherwise be avoided. To select a proper pump for a particular application it is necessary to utilize both system and pump performance curves. Figure IV-22 shows a system curve and pump curve.

A centrifugal pump (performance) curve is simply a tool which enables anyone to literally see how a pump will perform in terms of head and flow; whereas, a system curve is based on the total pressure a pump must produce to move the fluid through the piping system. The total

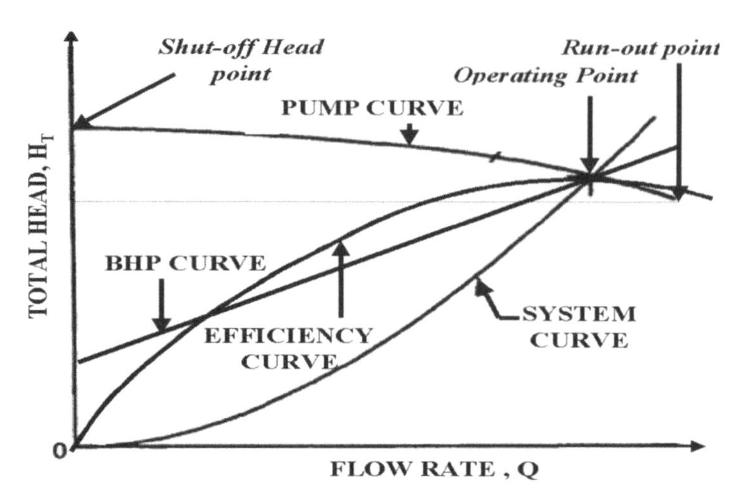

Figure IV-22
Pump and System Curve Chart

pressure of the piping system is the sum of the pressure due to inlet and outlet conditions and the pressure loss due to friction. In a piping system, pressure loss due to friction increases with increasing fluid flow; thus, system curves have positive slopes on pump performance charts. The operating point of a pump is determined by the intersection of the pump and system curves. The inlet/outlet pressure that the pump must overcome is the sum of the static, velocity and elevation pressures between the inlet and outlet of the piping system.

> **Note**: Pump performance and system curves can illustrate the basic interaction in the total system. They consist of a system curve showing the head required to pass a given flow rate through the piping system and a pump curve superimposed on the system curve. The point where the system and pump curves intersect is the balanced flow rate through the pump. In the absence of control valves, the system will operate at the intersection of the pump and system curves.

Many design elements affect the shape of the pump curve, and most of these cannot be changed by the user. As a result, centrifugal pumps are usually selected from the manufacturer's available designs to match the system requirements. Characteristics that can be changed by users to change the pump (performance) curve are the impeller diameter and the rotational speed. The pump curve change will cause the pump curve to intersect the system curve at a different rate of flow. A pump curve describes the operation of a pump for a range of flows at a defined speed. When selected properly, the pump will operate near its best efficiency point (BEP). This relationship of speed change or diameter change is often referred to as the pump affinity rules. Pump efficiency is greatest when the largest possible impeller is installed in the pump casing. Efficiency decreases when smaller impellers are installed because of the increased amount of fluid that slips through the space between the tips of the impeller blades and the pump casing. Efficiency also decreases as the rotational speed of a pump is reduced. However, the magnitude of the decrease in pump efficiency depends on the individual pump. See Figure IV-23.

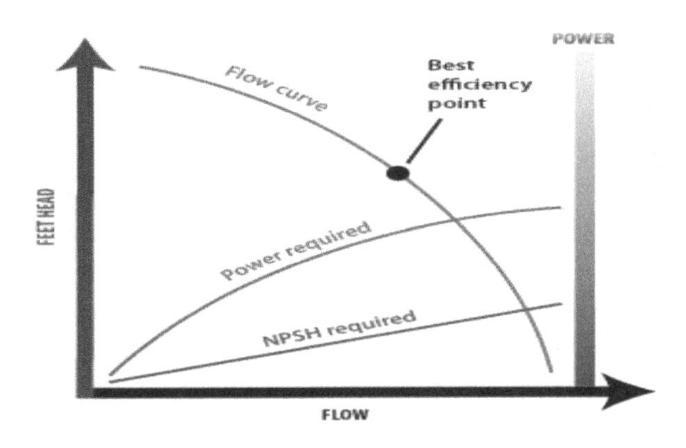

Figure IV-23
Curve Chart Showing Best Efficiency Point

Pump Terms

Head

Centrifugal pump curves show pressure as head, which is the equivalent height of water with (specific gravity) S.G. = 1. This makes allowance for specific gravity variations in the pressure-to-head conversion to cater for higher power requirements. Positive displacement pumps use pressure (psi) and then multiply power requirements by the S.G.

Static Head

The vertical height difference from surface of water source to centerline of impeller is termed as static suction head or suction lift (suction lift can also mean total suction head). The vertical height difference from centerline of impeller to discharge point is termed as discharge static head. The vertical height difference from surface of water source to discharge point is termed as total static head.

Total Head/Total Dynamic Head

Total height difference (total static head) plus friction losses and demand pressure from nozzles etc. (i.e. Total Suction Head plus Total Discharge Head = Total Dynamic Head).

NPSH

Net positive suction head is related to how much suction lift a pump can achieve by creating a partial vacuum. Atmospheric pressure then pushes liquid into the pump.

Specific Gravity

Weight of liquid in comparison to water at approximately 20°C (centigrade).

Specific Speed

A number which is the function of pump flow, head, efficiency etc. Not used in day-to-day pump selection, but very useful as pumps with similar specific speed will have similarly shaped curves, similar efficiency/NPSH/solids handling characteristics.

Vapor Pressure

If the vapor pressure of a liquid is greater than the surrounding air pressure, the liquid will boil.

Viscosity

A measure of a liquid's resistance to flow (how thick it is). The viscosity determines the type of pump used and the speed it can run, and with gear pumps, the internal clearances required.

Friction Loss

The amount of pressure/head required to force liquid through pipe and fittings.

Reading Pump Curves

When reading a pump curve, all curves are based upon the principle of plotting data using the x and y axis. With this in mind, the curves typically plotted are head vs. capacity, power input vs. capacity, and pump efficiency vs. capacity. Therefore, the constant between each curve is the capacity or x-axis. To determine the performance data at a particular point, first locate the operating point of the pump. This is the point where the system head curve crosses the pump's head vs. capacity curve. From this point move horizontally to the left until you intersect the y-axis. This will give you the head at which the pump will operate. Next go back to the operating point. By moving vertically down to the x-axis, you can find the capacity that the pump will operate. Now, at the determined flow rate, moving vertically to the input power curve intersection, move horizontally to the kW input y-axis where the appropriate value for motor input can be read.

REFRIGERATION CYCLE

Refrigeration is a process that removes heat from areas where it isn't wanted and transfers it to an area where it is unobjectionable. Refrigeration (and air conditioning) systems are divided into high- and low-pressure sides. The high-pressure side initiates at the condenser unit, and the low-pressure side initiates at the evaporator. The dividing point between high and low pressure cut through the compressor and the expansion valve. Air conditioning and refrigeration units work by manipulating refrigerant through refrigerant tubing interconnecting four main system components. These four components are divided into two different pressure areas across their systems: high pressure and low pressure. The four components (illustrated in Figure IV-24) are the compressor (1), the condenser (2), the expansion device (3) and the evaporator (4).

Compressor

The compressor is the heart of the system; it keeps the refrigerant flowing through the system at specific rates of flow, and at specific pressures. The rate of flow through the system will depend on the size of the unit and the operating pressures will depend on the refrigerant being used and the desired evaporator temperature. It takes refrigerant vapor

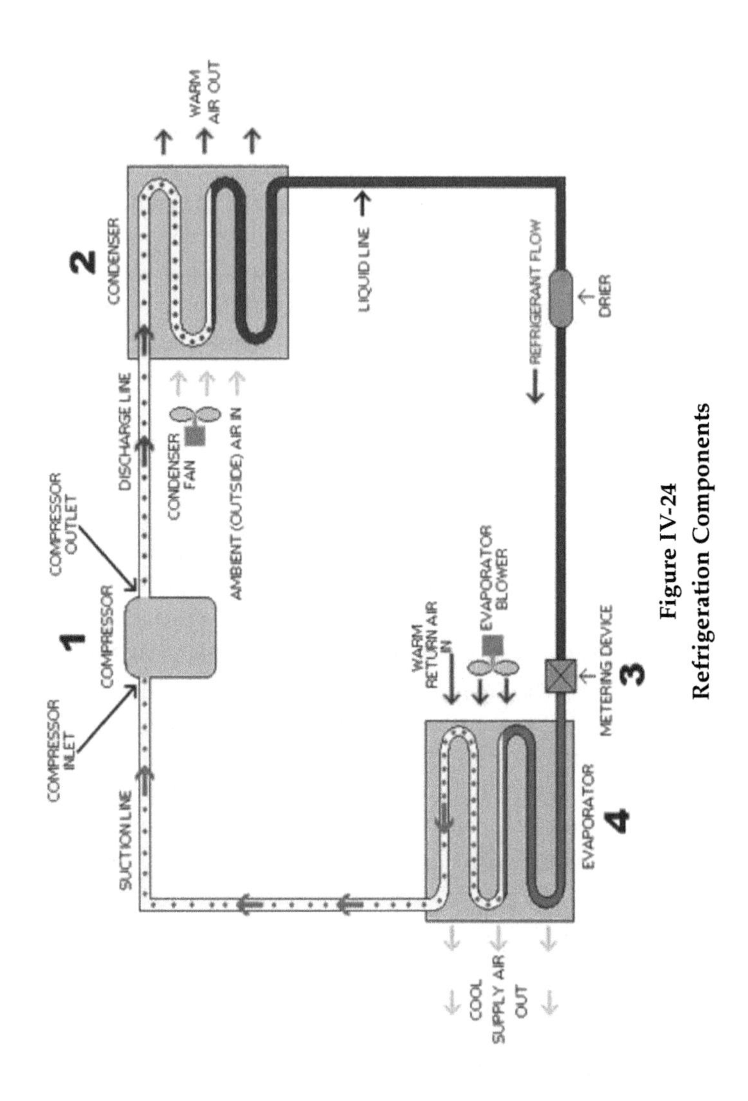

Figure IV-24
Refrigeration Components

in from the low pressure side (suction side) of the circuit and discharges it at a much higher pressure into the high pressure side (discharge side) of the circuit.

Condenser

The condenser is a heat exchanger. It is the condenser's job to dissipate the heat gain as a consequence of the refrigeration effect, as well as the heat of compression (occurring in the compressor). When the hot refrigerant vapor discharged from the compressor travels through the condenser, the cool air or water flowing through the condenser coil absorbs enough heat from the vapor to cause it to condense. The refrigerant is condensed at a relatively high temperature so that the air or water flowing through the condenser will be very cold relative to the temperature of the discharge vapor, allowing the heat energy in the vapor to move into that relatively cold air or water, thereby causing the condensing action. The dots inside the piping represent high-pressure vapor; the solid red color represents high-pressure liquid refrigerant.

Metering Device

The metering device (expansion valve) is the dividing point between the high pressure and low pressure sides of the system and is designed to maintain a specific rate of flow of refrigerant into the low side of the system. At this stage, high pressure liquid refrigerant will flow down the liquid line, through a filter drier that is designed to prevent contaminants from flowing through the system, and on to the metering device. When the high pressure liquid refrigerant passes through the metering device, its pressure will drop to a low pressure. It starts evaporating immediately!

Evaporator

The evaporator is a heat exchanger. It is responsible for extracting heat from the medium being cooled (air, water or other) and boiling the entire amount of liquid refrigerant to vapor (prior to entering the compressor). Relatively warm air (or water) flows over the evaporator coil. The refrigerant will evaporate in the evaporator (coil) at a temperature that's about 10° to 15° below the temperature setting if it's a refrigerator or freezer, and the temperature will drop to around 0° in the evaporator of an ice machine. The heat in the relatively warm air or water flowing over the evaporator coil will be absorbed into the cold evaporating refrigerant until it reaches the design setpoint or thermostat setting. The dots inside the piping represent low-pressure vapor; the solid blue color represents low-pressure liquid refrigerant.

REFRIGERATION GAUGES

Refrigerant gauge manifold sets are used to test the operating condition and cooling status of freezers, box refrigerators, air-conditioners, walk-in coolers, and other systems operated with various types of refrigerant. Problem diagnosis is accomplished by reading the amount of liquid or refrigerant vapor pressure within the system. Refrigeration mechanics should be thoroughly trained in the safe use and handling of refrigerant (types) and adequately trained in the use of these gauges for testing for leaks and charging and releasing refrigerants from these cooling devices. Adding and removing refrigerant should only be done by qualified technicians. See Figure IV-25.

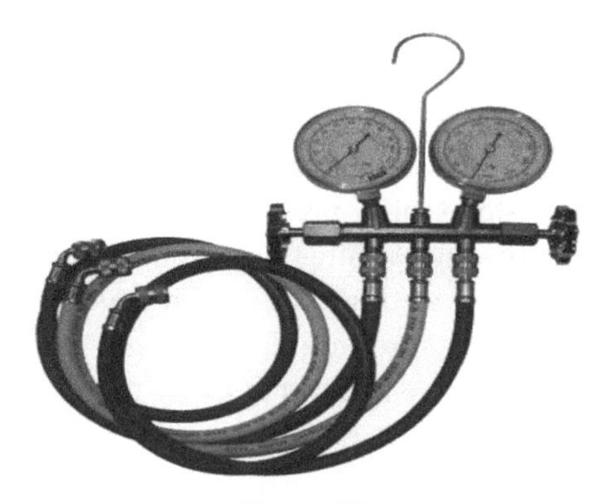

Figure IV-25
Refrigerant Gauges

Construction

A refrigerant manifold gauge set consists of three hoses, each with an identifying color, two gauges (measuring pounds per square inch, or psi) and control valves for manipulating the refrigerant. The red hose is the high-pressure service port, and the blue hose is the lower-pressure service port. The yellow hose is attached to the refrigerant cylinder or vacuum pump if the refrigeration unit is being charged or evacuated of air. The gauge colors correspond to the blue and red hoses. The blue (low-pressure) gauge (compound gauge) ranges from minus 30 to 20 and measures pressure in psi gauge (psig) or vacuum in inches of mercury. The red (high-pressure) gauge ranges from 0 to 500 and measures only pressure. When units are off, both gauges should register approximately the same number (static pressure). Static pressure readings determine if liquid refrigerant is present. See Figure IV-26.

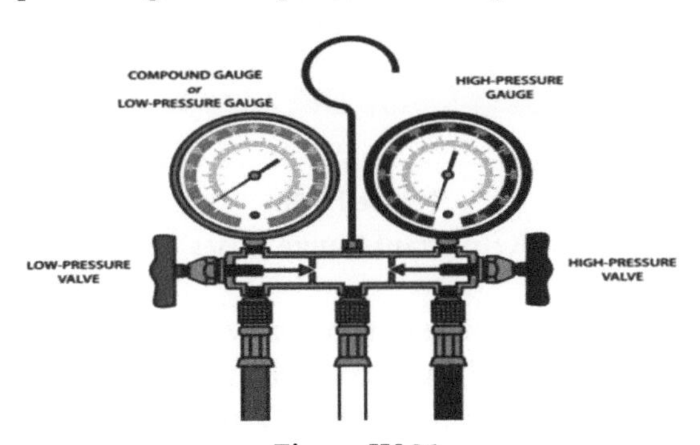

Figure IV-26
Vacuum/Refrigerant Hose

The center service port on the gauge set is connected to a refrigerant canister, a charging device, or an evacuator pump, depending on what the HVAC technician needs to do. Depending on which valves are opened or closed, the gauge set permits charging the refrigerant system on either the high side or the low side. See Figure IV-27.

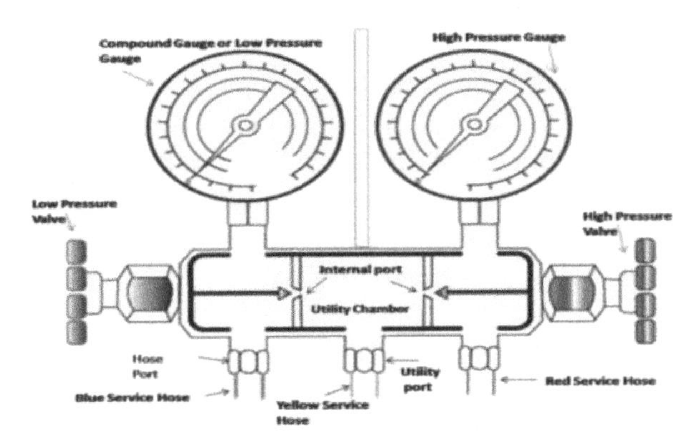

Figure IV-27
Gauge Internals

The threaded hose ends are connected to strategic points within the systems, giving the HVAC technician the ability to read refrigerant gas pressures and add or replace refrigerant. See Figure IV-28.

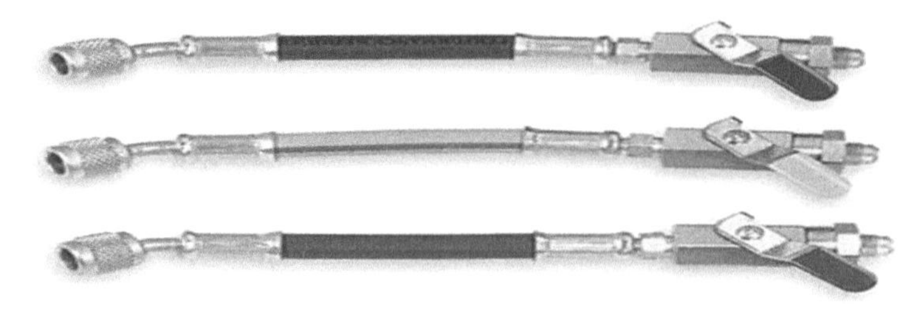

Figure IV-28
Refrigerant Gauge Hoses

Note: Some hoses come equipped with their own individual valves; in addition to the two control valves. The advantage of this arrangement is that it allows the HVAC technician to connect to the refrigerant cylinder and vacuum pump without disconnecting the service hose (utility hose).

Hook-up

Figure IV-29 shows the gauges, the refrigerant recharging cylinder and the vacuum pump hooked to an outside unit at the access ports. Note that the low side gauge (blue hose) will attach to the suction side of the system which is the lower-pressure side. The high-pressure gauge (red hose) will attach to the service port of the high side liquid refrigerant valve. Both of the service valves will likely have plunger pins inside them, and you will not have to turn any valves or valve stems to access the pressure inside.

Some systems may contain liquid refrigerant with pressures as high as 500 pounds. Use caution as you install your manifold gauges. Hose connectors are designed to spray the spillage away from your hand, but it can still be dangerous.

Service Ports

To keep moisture out of a refrigeration system, in addition to finding and fixing leaks, we need to know how to properly use a refrigerant gauge set with charging lines, and how to use cap-off plugs on the charging fittings. Central air conditioning systems, heat pumps,

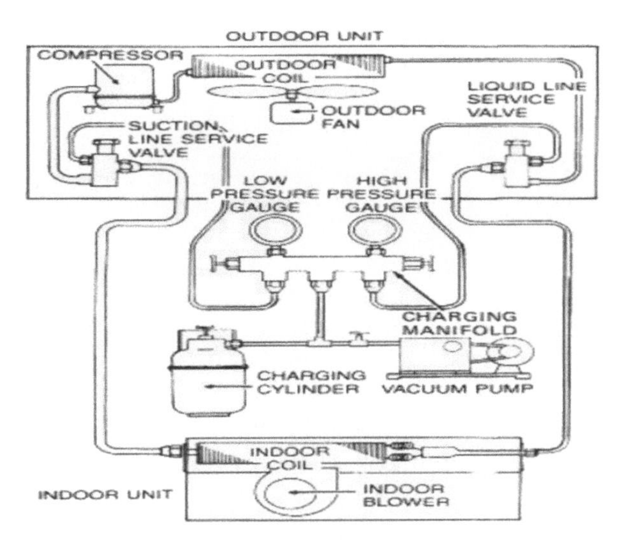

Figure IV-29
Refrigerant Gauge Hook-up

and split systems typically have service ports installed specifically for the attachment of test gauges for system inspection, evacuation, and charging. See Figure IV-30. Residential refrigerators, freezers, and window or portable air conditioners typically will not have these service ports. To service one of those latter devices you'll need to cut the refrigerant line and install (solder in place) a tee and a service port. Vampire taps and other piercing valves are available in various sizes to allow the HVAC technician to tap into the refrigerant lines on a system to perform diagnosis where service valves are not already installed. When connecting an HVAC refrigeration gauge set to test fittings on an air conditioner or heat pump we must:

- Connect the gauge set center supply tube to a canister of the proper refrigerant gas matching the refrigerant in the system being tested.

- Leave some positive pressure of refrigerant gases in each of the two gauge test connection hoses (the high-pressure side and the low-pressure side) so that when the gauge hose fitting is connected to the service port on the HVAC equipment, no outside air or moisture are pushed into the system piping.

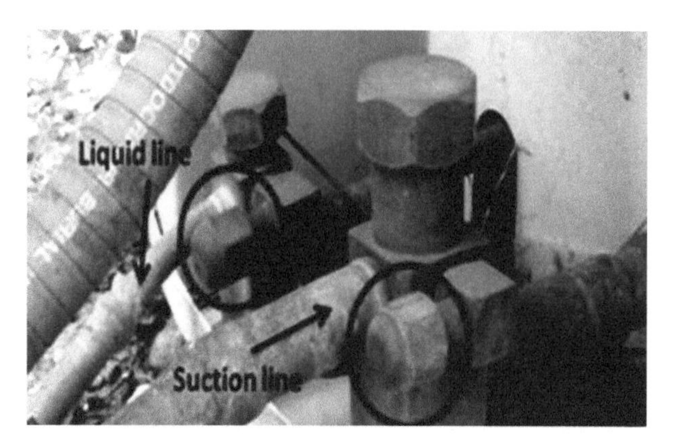

Figure IV-30
Refrigerant Service Ports

> **WARNING**! Do not send liquid refrigerant into the low side of a refrigeration system. Liquid refrigerant will enter the bottom of the compressor motor and can damage the compressor. Even if the compressor tolerates and passes the refrigerant through its pumping system, the refrigerant can carry away the lubricating oil from the compressor, and/or cause an air trap in the system.

PT Charts

HVAC technicians use refrigerant pressure-temperature (PT) charts to check the proper operation of their refrigeration systems. Pressure-temperature (PT) charts are valuable tools that are most often used to set a coil pressure so that the refrigerant produces the desired temperature, to check the amount of superheat above the saturated vapor condition at the outlet of the evaporator and to check the amount of sub-cooling below the saturated liquid condition at the end of the condenser. PT charts allow technicians to determine the amount of pressure they must set up to produce the desired temperature. By comparing pressure and temperature readings derived from the "refrigeration gauges" to the PT chart, technicians can find, repair or adjust system problems manifested by the equipment they are servicing. The process for doing so is four-fold: to identify the type and model of the refrigerant used in the unit, to obtain the PT chart for the model of your refrigerant, to choose a temperature from the left column of the chart then go to the right in the table row to see what the amount of pressure must be at that temperature, and to compare the value of the pressure from the

PT chart with your current gauge readings to determine the system's condition. See Figure IV-31.

Figure IV-31
PT Chart—Temperature • Pressure Chart

Temp.		Refrigerant							
°F	°C	HFC 410A	HCFC 22	HFC 407C	HFC 134a	CFC 12	CFC 502	HFC 404A	HFC 507
33	0.6	103.3	58.8	53.4	28.6	30.8	68.4	73.9	77.4
34	1.1	105.4	60.2	54.8	29.5	31.7	69.9	75.5	79.0
35	1.7	107.5	61.5	56.1	30.4	32.5	71.4	77.1	80.7
36	2.2	109.7	62.9	57.5	31.3	33.4	72.8	78.7	82.3
37	2.8	111.9	64.3	58.9	32.2	34.2	74.3	80.3	84.0

SHAFT ALIGNMENT

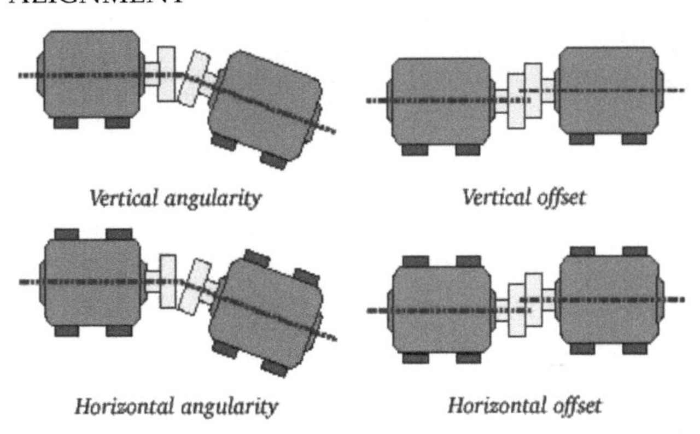

Figure IV-32
Types of Shaft Misalignment

Angular Misalignment

Also known as gap or face misalignment, angular misalignment is the difference in the slope of one shaft, usually the moveable machine, as compared to the slope of the shaft of the other machine, usually the stationary machine. Angular misalignment occurs when the motor is set at an angle to the driven equipment. The angle or mismatch can be to the left or the right, or above or below. If the centerlines of the motor and the driven equipment shafts were to be extended, they would cross each other, rather than superimpose or run along a common centerline. Angular misalignment can cause severe damage to the driven equipment and the motor.

Offset Misalignment

Also known as parallel misalignment, offset misalignment is the distance between the shaft centers of rotation measured at the plane of power transmission, typically measured at the coupling center. Parallel misalignment occurs when the two shaft centerlines are parallel, but not in the same line. They are offset horizontally or vertically (or both), displaced to the left or right, or positioned at different elevations.

> **Note**: There are also two planes of potential misalignment—the horizontal plane (side-to-side) and the vertical plane (up-and-down). Each alignment plane has offset and angular components, so there are actually four alignment parameters to be measured and corrected. They are horizontal angularity (HA), horizontal offset (HO), vertical angularity (VA), and vertical offset (VO). Combination misalignment occurs when the motor shaft suffers from angular misalignment in addition to parallel misalignment.

Couplings

Larger motors are usually directly coupled to their loads with rigid or flexible couplings. Rigid couplings do not compensate for any motor-to-driven-equipment misalignment, while flexible couplings tolerate small amounts of misalignment. Flexible couplings can also reduce vibration transmitted from one piece of equipment to another, and some can insulate the driven equipment shaft against stray electrical currents. Flexible couplings should not be used for the sole purpose of correcting excessive misalignment, as flexing of the coupling and of the shaft will impose forces on the motor and driven-equipment bearings. Effects of these forces include premature bearing, seal, or coupling failures, shaft breaking or cracking, and excessive radial and axial vibrations. Secondary effects include loosening of foundation bolts, and loose or broken coupling bolts. Operating life is shortened whenever shafts are misaligned.

Symptoms

Misalignment is not easy to detect on machinery that is running. Consequently, what we actually see are the secondary effects of these forces. Machines will exhibit symptoms such as premature bearing, seal, shaft, or coupling failures, excessive radial and axial vibration, high casing temperatures at or near the bearings or high discharge oil temperatures, excessive amounts of oil leakage at the bearing seals, loose foundation bolts, broken coupling bolts, unusually high numbers of coupling failures, cracked or broken shafts and excessive amounts of grease (or oil) on the inside of the coupling guard.

Alignment Process

Misalignment produces excessive vibration, noise, coupling and bearing temperature increases, and premature bearing or coupling failure. Shaft alignment should be an organized and simple process. It is important to establish a set procedure to be followed for every alignment from beginning to end. After alignments are performed over a period of time using the same procedure, less time will be required to perform subsequent alignments. All equipment should be locked and tagged out before any work is performed on it and machines given ample time to cool. The work area should be free of obstructions and debris. All established safety procedures in place at the facility should be followed. The following items should form part of any good alignment process or plan:

- Check shaft alignment of all critical equipment annually.
- Monitor vibration as an indication of misalignment. Misalignment might be caused by foundation settling, insufficient bolt tightening, or coupling faults.
- After 3-6 months of operation, recheck newly installed equipment for alignment changes due to foundation settling.
- Predictive maintenance techniques, including vibration tests and frequency spectrum analysis, can be used to distinguish between bearing wear, shaft misalignment or electrically caused vibrations.

WATER HAMMER AND CAVITATION

Water hammer is a (noisy) destructive force that can damage commercial plumbing systems. Sounds emanating from water system piping such as knocks, clunks, clangs and bangs are usually caused by a phenomenon called water hammer. Water hammer (hydraulic shock) occurs when you shut off the water suddenly, and the fast-moving water rushing through the pipe is brought to a quick halt (when a pump starts or stops or when a valve closes or opens), creating a shock wave and a hammering noise. Shock forces due to water hammer can rupture copper supply lines or cause leaking joints. Technically, it results from a series of pressure pulsations of varying magnitude above and below the normal pressure of water in the pipe. The amplitude and period of the pulsation depend on the velocity of the water as well as the size, length, and material of the pipe. Shock loading from these pulsations occurs when any moving liquid is stopped in a short time. See Figure IV-33.

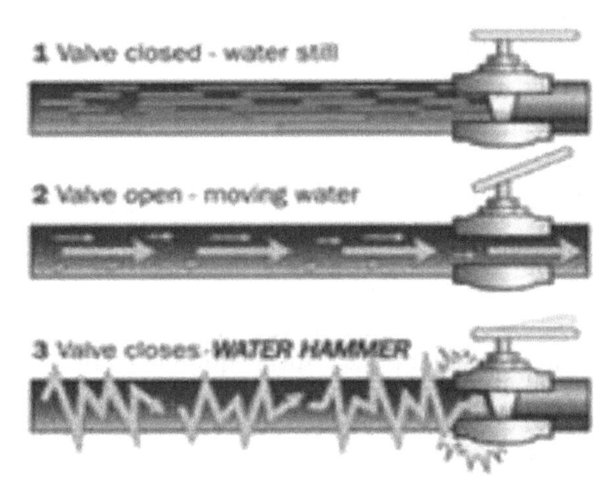

Figure IV-33
Water Hammer Action

> **Note**: It is important to avoid quickly closing valves in an HVAC system to minimize the occurrence of water hammer. **Also**: Water hammer doesn't accompany flowing water, so if your sound occurs while water is running, it isn't water hammer.

Cavitation

Like water hammer, cavitation is a significant cause of damage in pumping systems. It occurs when the pressure of a flowing fluid drops below the vapor pressure of that fluid. Technically, it is the formation and then immediate implosion of cavities in liquid—i.e., small liquid-free zones ("bubbles")—that are the consequence of forces acting upon the liquid. It usually occurs when a liquid is subjected to rapid changes of pressure that cause the formation of cavities where the pressure is relatively low. When entering high pressure areas, cavitation bubbles that implode on a metal surface cause cyclic stress. This results in surface fatigue of the metal causing wear. The most common examples of this kind of wear are pump impellers and bends when a sudden change in the direction of liquid occurs. See Figure IV-34.

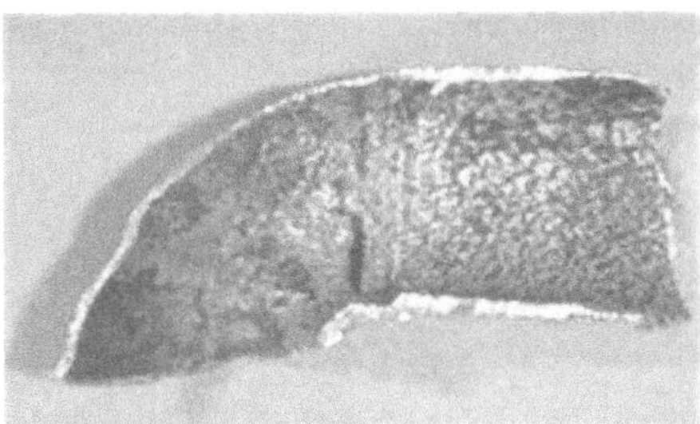

Figure IV-34
Common Areas Affected by Cavitation

Both water hammer and cavitation can occur in pumped fluid (water) mains. If no specific precautions are taken, the water hammer caused by pump startup or shutdown may cause the pressure to fall below the vapor pressure of the fluid. If this happens the continuity of the fluid is broken, or in other words the fluid is boiling (this is cavitation). Often the fluid contains dissolved gasses (water is often highly saturated with atmospheric air). If the pressure reduces because of water hammer, the fluid can become supersaturated with air, which is then released as small bubbles. The small bubbles then accumulate and form large bubbles. Since such accumulations cannot be dissolved as soon as they are released, the effect of water hammer can result in a net creation of air or gas pockets.

Cause and Effect

Water Hammer

Pumps, valves, faucets, toilets, and fast solenoid-activated valves are all examples of devices that can induce water hammer within a typical plumbing system. Water hammer can result in noisy, banging sounds as pipes rattle and expand to absorb pressure waves. Water hammer problems are generally traceable to inadequate system design and installation, including excessive system water pressure, lack of pressure regulation, improperly sized supply lines, inadequate support of pipes, long straight runs with no bends, and lack of a dampening system. Systems that are not protected against the ravages of water hammer can incur pipe vibration and noise, ruptured piping, weakened or leaking connections, damaged valves, corrupted meters, incorrect reading gauges, broken pressure regulators, loosened pipe hangers and supports, and ruptured pipes and tanks. Repairing any of these conditions (after the fact) is more expensive and inconvenient than designing a system right from the start.

Cavitation

Cavitation means that cavities are forming in the liquid that we are pumping. The cavities form for four basic reasons (all classified as cavitation, by definition). Each of them is actually a particular malady that requires specific corrective actions. Incidences include vaporization, air ingestion, internal recirculation and flow turbulence. When these cavities form at the suction end of the pump, several things happen all at once: a loss in capacity, the inability to build the head pressure, a drop in pumping efficiency and the collapsing of bubbles when they pass into the higher regions of pressure causing noise, vibration, and damage to many of the components. In each incidence, correction of these problems requires a specific cure.

Problem Resolutions (Water Hammer)

The most common solution to water hammering has been to install pipe risers inside the wall at each faucet or valve junction. Sometimes these risers would be as high as 24" or more, depending on the pipe diameter. In theory, the risers would trap air as the plumbing system is activated. The column of air acts as a natural damper, compressing as it absorbs residual shock waves from a sudden change in the supply flow. An air chamber will not drain properly if it is clogged. Remove its cap and ream out the accumulated scale inside the chamber.

Air Risers

Cushion pipes (air risers) eventually fail due to water logging. Over time, the trapped air in the risers dissolves into the water supply itself, and the water level gradually rises until the air chamber is completely void of air at all. With the use of "traditional" cushion pipes, the only real solution is to completely drain the water supply system at the lowest point and gradually re-pressurize it. This solution is temporary at best, since the air chambers will eventually become waterlogged once again, thereby eliminating their effectiveness. A more effective approach includes a valving arrangement such as represented in Figure IV-34 (right). Studies have also found possible health problems associated with air risers, such as an accumulation of rancid water, bacteria, minerals, and other muck that festers in the dark, dead-end chambers. If left unchecked and untreated, this could eventually contaminate the entire water supply causing unexplained illness. In modern practice, many plumbing codes now prohibit such air chambers in new construction. Rather than use air chambers to mitigate water hammer problems, design the system right from the start, and you'll never have to worry about it again. A combination of proper pipe sizing and water hammer arresters are all that's necessary in most situations. Other means for mitigating water hammer include:

Installation of Low-flow Fixtures

As water hammer is correlated to total flow velocity, one solution is to replace existing fixtures with low-flow types or water restrictors. Replacing tub and lavatory faucets or using flow restrictors could be enough to minimize water hammer.

Check Washers, Valves and Toilets

Another tactic is isolating the fixture that has the problem, inspecting the valve mechanism and replacing faulty washers in faucets or supply shut-off valves.

Secure Loose Pipes

Where accessible, tightly secure all pipes at frequent intervals to minimize rattling against hard surfaces. **Note**: While strapping will help to alleviate some of the rattling sounds resulting from water hammer, it does not address the water hammer condition itself.

Install Pressure Regulators

If the water pressure is high (in excess of 60-80 psi), adding a pressure regulator could help reduce water hammer. By reducing the pressure, you can retain sufficient pressure while reducing the likelihood of damaging water hammer. Install in-wall water hammer arresters; water hammer arresters help to absorb the pressure shock wave, virtually eliminating water hammer. Manufacturers make compact water hammer arresters for residential and commercial applications. For optimum performance, water hammer arresters should be installed in-line with the branch supply at the fixture tee with no intermediate bends. Furthermore, sizing must be done by a qualified professional working with the manufacturer. Failure to properly size the water hammer arrester will result in sub-optimal performance and wasted effort. It is a cost-effective solution for a single valve, on a single device. Versions are available that either screw in or can be sweat soldered.

Usually damage is limited to breakage of pipes or appendages, but water hammer has caused accidents and fatalities. Operating engineers should always assess the risk of a pipeline burst. The following actions may reduce or eliminate water hammer: reduce water supply pressure, lower system fluid velocities, shorten lengths of straight pipe, fit slowly-closing valves, install station bypasses, install pump flywheels, arrange piping in loops, shorten branch pipe lengths, improve pipeline control and install water hammer arrestors.

Problem Resolutions (Cavitation)

All pumps require well-developed inlet flows to meet their potential. Preventing cavitation can be done by decreasing pressure at the intake or increasing pressure at the discharge as well as reducing the sources of potential bubbles. When poorly developed flow enters the pump impeller, it strikes the vanes and is unable to follow the impeller passage. The liquid then separates from the vanes causing mechanical problems due to cavitation, vibration and performance problems due to turbulence, and poor filling of the impeller. This results in premature seal, bearing and impeller failure, high maintenance costs, high power consumption, and less-than-specified head and/or flow. To have a well-developed flow pattern, pump manufacturer's manuals recommend about 10 diameters of straight pipe run upstream of the pump inlet flange. Unfortunately, plant personnel must contend with space and equipment layout constraints and usually cannot comply with this recommendation. Instead, it is common to use an elbow close-coupled to the pump suction which creates a poorly developed flow pattern at the pump suction. Cavitation in pumps may occur in two different forms, suction cavitation and discharge cavitation.

Suction Cavitation

Suction cavitation occurs when the pump suction is under a low-pressure/high-vacuum condition where the liquid turns into a vapor at the eye of the pump impeller. This vapor is carried over to the discharge side of the pump, where it no longer sees vacuum and is compressed back into a liquid by the discharge pressure. This imploding action occurs violently and attacks the face of the impeller. An impeller that has been operating under a suction cavitation condition can have large chunks of material removed from its face or very small bits of material removed, causing the impeller to look sponge-like. Both cases will cause premature failure of the pump, often due to bearing failure. Suction cavitation is often identified by a sound like gravel or marbles in the pump casing. See Figure IV-35.

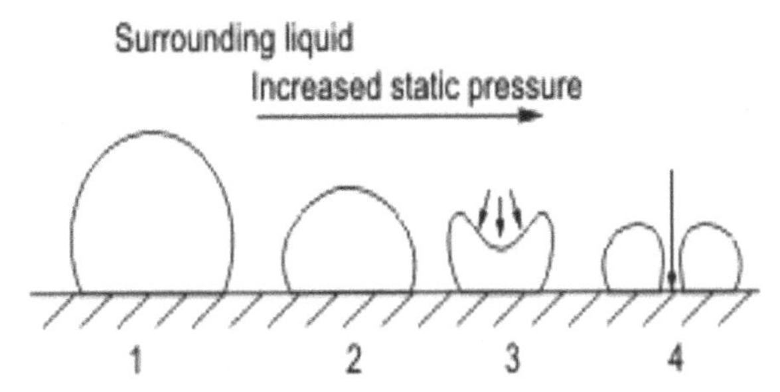

Figure IV-35
How Cavitation Occurs

Discharge Cavitation

Discharge cavitation occurs when the pump discharge pressure is extremely high, normally occurring in a pump that is running at less than 10% of its best efficiency point. The high discharge pressure causes the majority of the fluid to circulate inside the pump instead of being allowed to flow out the discharge line. As the liquid flows around the impeller, it must pass through the small clearance between the impeller and the pump housing at extremely high velocity. This velocity causes a vacuum to develop at the housing wall (similar to what occurs in a venturi), which turns the liquid into a vapor. A pump that has been operating under these conditions shows premature wear of the impeller vane tips and the pump housing. In addition, due to the high pressure conditions, premature failure of the pump's mechanical seal and bearings can be expected. Under extreme conditions, this can break the impeller shaft.

WATER TREATMENT

In the physical plant, water comes in two varieties, potable and industrial. Potable water is intended for human consumption, especially when free of harmful contents such as industrial waste, chemicals or animal waste. Typical uses (for other than potable purposes) include toilet flushing, washing and landscape irrigation. Industrial water is never consumed or used under situations that require a high degree of sanitation. Industrial water refers to the water used in power generation, heating, air conditioning, refrigeration, cooling, processing, and all other equipment and systems that require water for operation. Industrial water requires water preparation or chemical treatment, or both, according to the type of system in question, to avoid the problems. It is industrial water we will cover here. Examples of industrial water systems and their uses are:

Closed Water Systems

These include closed hot water, closed chilled water, and diesel jacket systems.

Cooling Water Systems

Those used in cooling towers, evaporative coolers, evaporative condensers and once-through systems. Applications are broad, ranging from simple refrigeration to temperature regulation of systems.

Steam Boiler Systems

Systems that include space and hot water heating, sterilization, humidification, indirect food processing, and power generation.

System Problems

Problems found in industrial water systems are attributable to reduced or restricted water flow or other changes in operational parameters, and often caused by corrosion, deposits, and biological growth.

These problems result in reduced system efficiency (higher operating costs), increased equipment replacement costs, and reduced safety. At times they can be serious enough to cause complete system shutdown. The problems in industrial water systems fall into three main categories:

- *Corrosion*—metal deterioration resulting from a refined metal's tendency to return to its original state (i.e., the ore from which the refined metal was produced).
- *Deposits*—composed of mineral scale, biological matter, and suspended or insoluble materials (e.g., sludge, dirt, or corrosion byproducts). Deposits can be created by the attachment of deposit-forming materials to pipe or equipment surfaces, or by settling and accumulation.
- *Biological*—microbiological organisms (algae, fungi, bacteria). Algae are microscopic plants that may grow in various industrial water systems but most commonly appear on the distribution decks of cooling towers. Fungi are living organisms that may cause damage to the wooden parts of cooling towers by causing decay. Slimes are accumulations of these biological contaminants that foul and corrode the cooling water equipment.

Treatment Objectives

Industrial water is treated to maintain component efficiencies, prolong usable life of systems and reduce repair frequencies. These objectives are met by treating the water to prevent scale and to control corrosion, fouling, and microbiological growth. To meet these objectives, an adequate and continuous supply of both properly conditioned makeup water and conditioned or chemically treated system water (i.e., water within the water-using system) is produced. Scale describes specific types of deposits caused when mineral salts, dissolved in water, are precipitated either because their solubility limits have been exceeded or as a result of reaction to water treatment chemicals. See Figure IV-36.

Scale adheres to pipe and equipment surfaces and its formation results in loss of heat transfer and restricted flow of water or steam. Many different types of scale reflect the quality and characteristics of the makeup water and the type of chemical treatment being applied. See Figure IV-37.

Suspended solids refers to any materials present in the water stream that are not actually dissolved in the water. They can result from

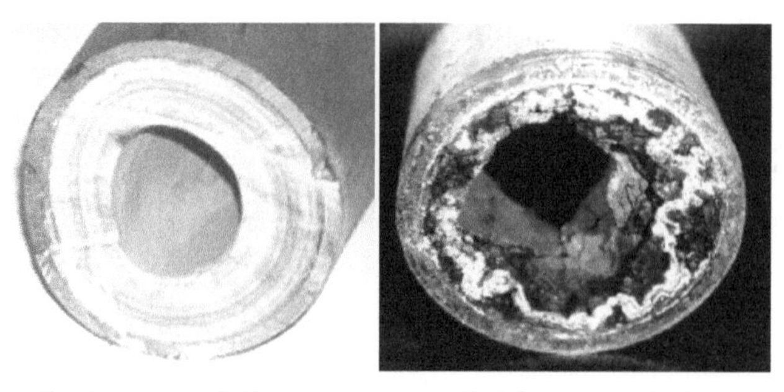

Scale accumulation **Solids accumulation**

Figure IV-36

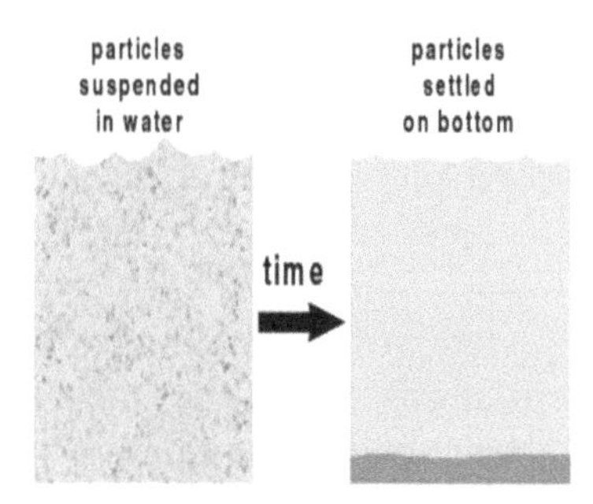

Figure IV-37
Settling of Suspended Solids

the presence of dirt, silt, and sand in the makeup water or can be introduced into the water from air in a cooling tower system. Biological matter, both dead and living, can be a form of suspended solids carried in the water stream. Corrosion products, such as iron oxide, are also forms of suspended solids that often originate in the system piping.

Water Treatment Methods

Both external and internal treatment methods may be used in a given system. External water treatment (pre-treatment) involves various processes to remove or reduce hardness, alkalinity, dissolved gases, or other impurities before the water enters the water system; i.e., steam boiler, cooling tower, closed hot water system or chilled water system. External treatment equipment processes and water treatment chemicals reduce or remove impurities contained in the makeup water before the impurities in the water stream enter the internal system. The most effective way to protect the system, reduce boiler problems, and improve operating efficiency is to use a process of removing impurities before they enter a system. Internal water treatment involves the treatment of water directly within the water system. Internal treatment of water is a process of adding chemicals to the system to control deposition and corrosion. Internal water treatment, together with proper blowdown control, controls the water impurities that have not been removed or reduced through external treatment. See Table IV-2.

Table IV-2
Impurity Removal Methods

Impurity	*Removal/Reduction Method*
Hardness (calcium and magnesium)	Sodium ion exchange Hydrogen ion exchange Lime-soda softening Evaporators RO Electrodialysis
Alkalinity (bicarbonate and carbonate)	Lime-soda softening Hydrogen ion exchange (followed by degasifying) Dealkalinization (chloride ion exchange)
SS/turbidity	Filtration/clarification
Dissolved solids	Demineralization (deionization) Evaporators RO Electrodialysis
Dissolved iron	Aeration (converts to precipitated iron), then filtration Sodium ion exchange (iron exchange (iron will foul the resin)
Dissolved gases (carbon dioxide, hydrogen sulfide, methane)	Aeration Degasifying

Index